Gentle Gorilla
The Story of Patty Cake

Gentle Gorilla
The Story of Patty Cake

by Susan Kohn Green

Richard Marek Publishers, New York

Printed in the United States of America

**Library of Congress Cataloging in
Publication Data**

Green, Susan.
  Gentle gorilla.

  1. Gorillas—Legends and stories. I. Title.
QL795.G7G73   599′.884   77-19009

ISBN 399-90004-7

CONTENTS

ACKNOWLEDGMENTS

I would like to thank many people:

My father Dr. Irving L. Kohn who gave me the wonder of life; for all my life I have heard him marvel at the existence of life itself.

My mother Naomi Kohn who shared her sense of grace and beauty with me and deliberately developed in me a knowledge that without wanting to be curious and adventuresome, without working toward some accomplishment which is and always will be part of you, life seems just that much less complete;

My sweet Elizabeth who, from the time she was born up to until she was sixteen months old, was happy to play next to me each morning while I worked, coming to me every now and then to play, lay her head on my lap and hug my leg while I leaned down to hug her; giving me the warmth I need. Without her sweet nature this book would never have been completed.

The great anatomy teacher, Robert Beverly Hale, who introduced me to the wonders of the living body, the perfection of the mechanism in which each part has its function in the performance of life. He taught us that there is infinite beauty in the reality of form;

The brilliant tiger tamer Gunther-Gabel-Williams who, through his energy, work and innate beauty, showed me that a work of beauty must be a universe within itself. When I watch Gunther work with his tigers, the world dissolves until nothing else exists but the fluidity of this incredible man and his cats;

All the keepers at the Central Park Zoo who invited me into the

Lion House and then accepted me as one of them; they made me feel more than welcome and are my friends;

John FitzGerald who allowed me to stay and watch Patty Cake, Lulu and Kongo, his abilities and personal work with the animals, and his love for them, I admire more than I can say. I will never be able to thank him enough.

And, of course, Richard Marek, whose first reaction to the book made me feel an incredulous joy which I will never, ever forget, and whose sensitive work on the book was invaluable.

for the animals

INTRODUCTION

As an art student I spent long hours at the Central Park Zoo studying and drawing the animals years before Patty's birth. And the keepers had come to know me.

"Your smile is your pass," Dick Berg, a keeper, had said when I showed him my official permission slip to enter the Lion House before it opened to the public each morning. By September, 1972, when Patty was born, the zoo had become a regular part of my life. And miraculously, it seemed to me, I had become an accepted part of the life of the zoo. After Patty was born and the public was barred from the Lion House, John FitzGerald, Director of the Zoo, allowed me to stay.

I have always loved animals. And now, standing by the cages with pad and pencil, I drew the animals as they were, in attitudes and situations that actually existed and took place again and again. Slowly I came to marvel not only at the gorillas' beauty, but at their intelligence and their capacity for emotions. I began to keep a detailed diary of the life in those cages. For the next year and a half the gorillas were my life.

CHAPTER 1
Birth
September 3, 1972

Birth is most often private. It is something that happens between the mother and child and is hardly noticed by the world. And so it was with Lulu and her baby, so private that even the pregnancy had gone unnoticed and the baby was a surprise.

It was late in the afternoon on a sun-flooded day early in September. The zoo in Central Park was crowded with couples who strolled and hardly noticed the animal cages, with the lonely who had come hoping to meet friends, and with romping, excited children whose parents had brought them on this last holiday before school opened in the fall.

The gorillas had spent the day in the outside cage. They had already had their afternoon baths and had splashed exuberantly in water that rushed from a hose. It was at the whim of the keeper that they had their welcomed shower, and he enjoyed it as much as they did, as he played the stream over their bodies. They slapped their chests with opened palms and shook their bodies, sending the water spraying over the watching crowd. Then it was over. The keeper turned off the water and wound up the hose. He had other things to do.

Now, quieter, the 350-pound blackback male, Kongo, stuck his tongue deep inside a paper cup. His sharp eyes watched the people

15

as he sucked on the chocolate ice and licked it slowly off his fingers. He ate it leisurely, finally chewing the cup itself, enjoying the remaining vestiges of sweetness. When he was finished, Kongo put his hand between the bars toward a small boy eating an ice-cream cone. The boy ogled him silently as he swirled the soft ice cream with his tongue, spiraling up to the top. Kongo watched with intent, avaricious eyes.

Kongo's mate, Lulu, did not notice. She lay quietly in the shade at the back of the cage, tapping the sides of her stomach with her fingers. She glanced up briefly as keeper Richie Regano offered her a pretzel. He held it out to her temptingly, but at the sound of his familiar voice she merely stared for an instant and then turned her head away toward the brick wall. Richie looked at her wondering-ly, hurriedly thrust the pretzel into Kongo's outstretched hand, and went back to his chores. It was nearly time for the animals to be put in the Lion House for the night. There was a great deal to do.

Inside the house Dick Berg was already cleaning the cages and distributing the evening food.

The long corridor was flanked by the rows of cages, and on the left, the first three belonged to the gorillas. Of these the first, the home cage, was the largest, eighteen feet by nine feet, double the length of each of the other two. The cages looked gray and bare, made of cement and steel, brick and tile, enclosed by a proscenium wall of bars. Yet the tall, arch-shaped windows above allowed sun-light in. Below the windows in each cage stretched a thick wooden plank which I called the balcony. The apes often sat there and stared into the outside world. They spent half their winter of confinement on those balconies.

All these cages were connected by heavy plate steel doors which were controlled by a pulley just outside the bars. It was in these cages that the gorillas Kongo and Lulu had lived nearly all their lives. They were now seven years old.

Outside, the clock above the archway next to the gorillas' cages began its chiming of "Three Blind Mice." People gathered there to watch the bronze animals begin their hourly dance around the clock.

With his audience temporarily lost to the fantasy animals, Kongo hauled himself up and reached lazily to tear a leaf from a branch. Lulu stirred. She took a few swings on the tire which hung in the

16

cage, then returned to her corner and lay down once again. It was just four o' clock.

The clock's melody slowed like a music box running down, then stopped. And the bronze animals once again stood motionlessly on tiptoe. As the spectators dispersed, some of them drifted back to the gorillas' cage. They turned their attention to Kongo, who put one massive hand between the bars to beg for peanuts. He caught one, cracked it open, picked the meat from his palm, then meticulously brushed the broken shell from his hand. As soon as he had popped the nut into his mouth, he put out his hand and demanded more. The crowd laughed.

One of the visitors, Mrs. Lenore Mintz, glanced at Lulu and noticed something odd about her. There seemed to be a little swollen pink protuberance on the ape's rear end.

That's strange, she thought. *I thought that only chimps and baboons had that bare red spot back there.* As she turned to mention the oddity to her husband, the ape suddenly rose, waddled awkwardly to the front of the cage, and squatted. As Mrs. Mintz caught her husband's sleeve, she heard someone behind her say, "Something's hanging from the ape."

And then: "Oh, my God! She's having a baby!"

In seconds, a tiny, still, wet gray infant gorilla slithered to the ground of rough cement. Just born.

Lulu bent over it curiously and sniffed. Her tongue flicked out of her mouth and brushed it; then she picked it up. Holding it to her face, the mother gorilla began to lick the slimy placenta from the hairless, inert form. In another moment she snipped the umbilical cord with her teeth. The baby hung limply in her hands.

The crowd began to stir.

"Where's a keeper? My God, someone get a keeper!"

"Didn't they know this was going to happen?"

"Oh, my God, look at it!"

They gaped in silent fascination as the slight female gorilla held her baby. She held it upside down at first, staring at it. She turned it to cradle it in her arms. Again she looked at it and licked its tiny face. Then she put it down onto her lower belly, holding it against the dark hair of her stomach with the palm of her hand. It disappeared under her.

Richie came running and vaulted the railing. As he peered excitedly in at Lulu, she reached out between the bars and tugged gently on his sleeve. With her other hand, she drew the newborn infant from under her belly. She held it out in her palm for him to see, then suddenly clasped it to her breast, enclosing it with her arms. She began to dart around the cage, shaking her free arm and her head in excitement, the infant pressed close against her belly. Then, just as suddenly, she sat down to look again at the strange creature. Again she licked its head.

Kongo sat on the other side of the cage, watching the new mother and child apprehensively. His arms were folded across his chest; his head was thrust slightly forward. He seemed restless, standing up, darting away from them, then turning to look again. He paced up and down alongside the bars, and as his agitation grew stronger, he bounded back and forth against them, faster and faster. He raced to the door, put his hands under it, and tried to lift it. He couldn't and became frantic, galloping first to the back, then to the front, throwing fearful glances toward his familiar companion and this "thing" that she had. He whirled back to the door again as if to escape from the strangeness of this unknown. The inside cage was his refuge.

In the meantime, Richie ran into the Lion House.

"Lulu has a baby!" he yelled at Dick Berg, and Dick, thinking that Lulu had grabbed some woman's child, dropped the tin pan of fruit that he was carrying. Apples and oranges spilled onto the cement floor as he went rushing out.

To find Lulu nuzzling the tiny gray thing. And the crowd applauding.

He took one look and turned around. And as he raced inside, he passed another keeper. "Someone call Fitz!" he yelled at him. "Hurry! Lulu has a baby!"

The keeper stopped dead still and stared.

Inside, Dick and Richie unlatched the chain that held the door in place and opened it. Kongo, who had been hovering frantically near the door, raced inside and into a corner, almost cowering there, as far from Lulu as possible. She followed him in, as she usually did, in her lopsided amble, but now she held a baby carefully against her soft body. That morning she had gone outside alone.

The crowd faced an empty cage. There was little to show that moments ago a baby had been born there.

18

CHAPTER 2
Patty Cake
First Month

John FitzGerald, director of the Central Park Zoo, was called away from a baptismal celebration. When he arrived at the Lion House, he found the keepers assembled in front of the cage and Lulu cradling the baby in her arms. Fitz stood hunched over the railing, watching in awe as Lulu cuddled the infant and licked it gently, turning it over and over in her hands, pressing the amazingly small fingers between her teeth to clean them. He leaned over the railing and watched the gentleness with which this young mother performed her first examinations, looking at and licking the baby's every part. He stood there for hours, observing with unobtrusive intensity that isolated him from the questioning hubbub of the others. The keepers eyed him curiously for a while but eventually went back to work. There was no clue to his private thoughts in the short, direct questions he asked them. It was only when Lulu finally nursed her baby for the first time the following day that he made up his own mind about his greatest concern: He decided to leave Lulu and the baby together.

A twenty-four-hour watch was set up immediately, and the Lion House was closed to the public. The vigil was maintained for thirty-one days: Often working sixteen hours a day, Richie Regano, Eddie Rodriguez, Luis Cerna and Raoul Ortiz watched for any sign that there was something amiss with the infant.

Of course, no one could keep the news of the birth from the press. This was the first birth of a gorilla in captivity in the area (it was incorrectly rumored, in the world), and within a day the New York *Times* asked for an interview. Under the stipulation that no cameras were to be allowed, Fitz decided to permit one reporter, Deirdre Carmody, to see the infant.

It was difficult for Deirdre to see much of the baby, for it was so small and so like its mother in coloring that it was hidden on Lulu's body. But she did see it nuzzling and the care which Lulu displayed as she briefly held the baby up against her cheek. Deirdre, too, was suddenly and totally entranced by the sight of this mother and child. Only when she was back in her office writing her article did she realize that she had neglected to ask Fitz one important question. She dialed his number.

"Fitz," she asked, "can you tell me what you are planning to name the baby?" Her pencil was poised over her note pad.

The day after the baby's birth, the proud and happy keepers had retired to the keepers' room next door to the home cage. After much unexplained off-key singing, they announced that they had named the baby Sonny Jim. Eddie Rodriguez's little black eyes gleamed in secret hilarity and enjoyment as he later told me of their decision.

"Uh, well, the keepers named it . . . you know; since they were there . . . they named it—" There was an awkward pause; then he seemed to plunge in, "They named it Sonny Jim."

"Oh, really?" Deirdre jotted it down. "What made them pick that particular name?"

"Gee, uh, I really don't know." There was another lag in the conversation.

"OK. Thanks a lot, Fitz." Deirdre turned back to her typewriter.

That evening her telephone rang at home. It was the city editor, and he was chuckling.

"It's a nice piece you wrote about that gorilla," he said, "but you sure as hell weren't brought up on the streets of New York, were you?"

"No," Deirdre was plainly puzzled. "What do you mean?"

"Well, there's an old bawdy ballad here that goes: 'Lulu had a baby/She named it Sonny Jim/She threw it in a piss pot/To teach it how to swim.' "

Deirdre gasped.

"Wait a minute; it gets worse." The man sang a choice verse or two.

No wonder Fitz was embarrassed, she thought and prepared herself for the worst. The following morning when her article, "Birth of Male Gorilla Is Pleasant Surprise for the Zoo," appeared, discreet and not-so-discreet humming of "Lulu Had a Baby" could be heard wherever she happened to be.

On September 12 Deirdre was again at the zoo to see Lulu and the baby. But this time she was one among dozens of newspaper reporters and television crews. The entire New York press had finally been invited to see Lulu and little Sonny Jim.

Among those assembled in front of the cages was Kenneth Youngstein, a primatology student who had been working on the mother/infant relationship of Macaque monkeys. Impulsively he had called the zoo, and to his tremendous surprise and gratification Fitz had invited him to see the family. By now he was recording the activities of the animals every day.

Jorg Hess, the famous Swiss primatologist, had been delivering a paper in the United States when the baby was born. He, too, came to New York to see the newest arrival.

Now Lulu and Sonny Jim were allowed into the outside cage in which the birth had taken place. More than 100 people, held back by police barriers, crowded silently in front of the cage. The television crews and newspaper people crowded into the space between the spectators and the cages. It was impossible to see anything but bobbing heads and shifting bodies as people struggled for a glimpse of the gorillas. Ken and I, acquaintances by that time, stood off to one side watching the melee when Dr. Hess arrived. He greeted Fitz and then joined us on the sidelines. He and Ken introduced themselves; I overheard them as they talked.

"Are you sure it's a male?" asked Dr. Hess, "Did you see the genitalia yourself?"

"No, they said it was a male, and I assumed that it was."

Most of the gorillas are missexed, you know. Achilla, the famous gorilla in Zurich, was originally thought to be a male. They named it Achillis. Now she's Achilla, of course. Sometimes the most learned men don't find out until years after the baby's birth. It can be most embarrassing." Leaning a pad of paper on the railing, Dr. Hess began to draw a diagram of the genitals of the male and female gorilla infants, surprisingly similar to the untrained eye. The two men bent over the drawing, discussing the possibilities.

"Of course," said Dr. Hess seriously, "there's one sure way you can tell. You see, the male will urinate this way. . . . " Again they leaned over the drawing. I heard Ken laugh.

Dr. Hess was asked to join Fitz for an interview. Ken headed back toward the cage and pushed his way to the front of the crowd. With a spyglass to his eye he waited for the baby to urinate.

Fitz was speaking with the press. At moments like these I got the impression that he was itching to get back to his real work, supervising the construction of the bird house interior or holding a sick animal as the vet examined it. He loved working directly with the animals more than anything else. But these interviews were a necessary and important part of this schedule, and he tolerated them with quiet good humor.

Fitz was surrounded by newsmen now; Ken dashed up to him and plucked at his sleeve.

"Fitz!" He almost danced in his excitement. "Fitz, it's really important. I think you'll want to hear this now!" The newsmen looked at them curiously, and the two men stepped aside. Ken began to explain, speaking rapidly, gesticulating toward the gorillas. For once Fitz's reserve was shattered. Even I, yards away, could hear his exclamation of surprise.

"What!"

Again Ken eagerly explained. The pantomime was repeated.

"You're sure?" Ken nodded. Then he showed Fitz the diagrams. Fitz took off his dark glasses, rubbed his eyes, and turned back to the press.

"Well," he said, "there's been a little change." A little smile crossed over his face. The press waited.

"I, uh. . . ." He paused. "I am happy to tell you that Sonny Jim is a . . . girl."

In the crowd of reporters that surrounded Fitz, Peter Coutros of

22

the *Daily News* was taking notes. It was he who later suggested that his newspaper run a contest to name the baby. The contest was announced in the *News* the next day, and New Yorkers responded. Suddenly thousands of people were pondering names appropriate for a baby gorilla. By the deadline, September 21st, 31,134 names had been suggested, among them Love Lumps, Mazel Tov, Lulu Belle, KuLu, and Patty Cake.

Patty Cake was the name which John O'Connor, a fireman with the New York City Fire Department, had sent in. His wife's name was Patty, and they had planned to name a daughter after her. But there had been three sons, and until a little girl came along, they thought they would let the little gorilla use the name.

Patty Cake was chosen the winner by Commissioner of Parks August Heckscher. When the little gorilla grew up and the full name was no longer appropriate for a 150-pound animal, "Patty" would remain.

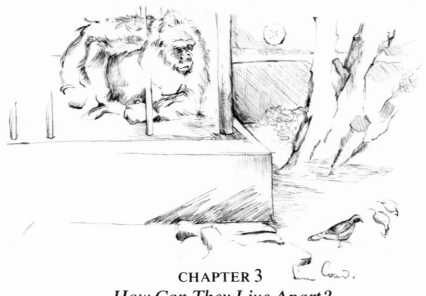

CHAPTER 3
How Can They Live Apart?
First Month

Kongo was separated from Lulu and Patty Cake the day she was born. At seven years he was still a "teenager" and had not yet settled down into the less playful attitudes of an adult gorilla. We wondered what his reaction to the infant would be once his initial apprehension had disappeared. Would he want to take possession of the infant? What would happen if he became jealous of Lulu's attentions to the helpless infant? What would happen to the baby if, in some sudden spurt of energy, Kongo stepped on her or hit her? Or if, in playing with Lulu, he rolled on top of her? Fitz, uncertain of the answers, decided to take no chances. When, soon after the birth, Kongo bounded into the adjoining home cage, the steel door lowered. He was alone.

Kongo and Lulu had never before been separated. They had been captured in the Cameroon, from different troops of lowland gorillas, when both were about six months old. From that moment they had lived together, first in Africa and then in Denmark at the home of the dealer to whom they were sold. They had even arrived at the New York airport together in the same crate. Fitz, then a keeper, had gone to pick them up, and when the nails had been removed from the wooden crate, and the lid lifted, two little gorilla heads had tilted up at him, and four arms had reached out to him. They were fourteen months old.

Fitz had tried to separate them once soon after their arrival at the zoo. Kongo had somehow cut himself; the gash on the side of his throat was deep and bleeding. Fitz had taken Kongo from the cage to take him to the ASPCA. Both animals had become hysterical, screaming and reaching for each other through the cage bars, Kongo struggling to wrestle himself free of the men who held him, and Lulu running back and forth. Finally, Fitz had opened the cage door once again, and Lulu had run to her companion, touching Kongo, clasping him. The apes had calmed down immediately. They had quietly walked to the car and had been driven to the ASPCA peering peacefully out the window at the passing city.

No one had tried to separate them again.

Until now.

At the beginning Lulu was completely caught up with the infant and with the attention that she, too, received. Lulu was watched and stroked and fed continually. She was becoming spoiled.

A few days after the baby's birth Lulu discovered that if the infant cried out, the keepers came running with cookies and fruit and Italian ices. She soon learned to take advantage of their concern. Every once in a while she would put the baby down and leave it, just, it seemed, to make it cry out. At the sight of the helpless infant the keepers, afraid of what Lulu might do if they ignored her demands, complied again and again. As soon as she got what she wanted, she picked up the baby and nestled her in her arms. Lulu put the infant down more and more often, and it soon became a continuous cycle of blackmail and compliance, as she earned rewards for her negative, unnatural behavior.

Ken was very disturbed by what he saw. He believed that if Kongo were introduced to the baby, he might very well act as a check on Lulu, his very presence a discipline. Ken and I agonized over the separation, convinced it would be better for all three if they remained together from the beginning of the infant's life. For now, Lulu not only was using the infant for her own purposes, but was at the same time creating a bond between herself and the infant that excluded Kongo completely. Once the relationship was established, it would be a difficult, perhaps an impossible readjustment later to include the male in the group.

Finally, at 9 P.M. on September 12, the evening of a press confer-

26

ence, as Ken was getting ready to go to a party, the telephone rang. Fitz was at the other end of the line.

"Lulu's in trouble," he said. "Can you come right away?"

Central Park was dark and deserted, but at the Lion House the lights were on. Fitz and the keepers stood helplessly in front of the gorilla's cage. The infant, alone in the middle of the floor, screamed. Lulu strolled around the cage, and the keeper on duty alternately scolded, teased, or cajoled her from the bars. The men stood against the railing opposite. Fitz slouched, his hands deep inside the pockets of his corduroy jacket. Worried, anxious voices rose in the empty house as Ken argued to introduce Kongo to the little group.

Finally, Fitz agreed.

The following morning it was a tired and apprehensive group that gathered. Eddie opened the door between the two cages. When Kongo entered, Lulu backed away from him, and the two adults sat down and faced each other. Then Kongo slowly walked to her and touched first her, then the baby, gently. The people as well as the gorillas began to relax.

Then Kongo gave his mate a back slap. It was a perfectly normal sign of salutation, but when the noise resounded from the cage, tension filled the long room again. In a flurry of movement the animals were separated. Once more the door clanged sharply against the cement floor between them.

Protesting, Ken turned to Fitz, but the director shook his head.

"He could hurt her," Fitz said sadly. "We just can't take the chance."

It was early fall; the weather was still warm, and the apes were allowed outside. A police barrier was set up to keep the crowds back, and Lulu and the baby were given the small cage next to Kongo.

Kongo! I had always known him as a young and rambunctious ape, chasing Lulu throughout the space of the cage, teasing and hitting her as he swung by on the tire, swinging and racing with the power and special energy of play. But now, separated from his mate, he became nervous and sullen. He would sit in his isolation, staring at his hands or out into the space beyond the cage. He would put a fallen autumn leaf between his teeth or manipulate it

27

with his lips, a sign often of discontent. He would sit still for long minutes, then suddenly charge across the cage to a spot near Lulu on the other side of the wall. He would stand there on his knuckles, his legs straight, completely alert for a few seconds. Then his huge, powerful body would seem to droop, and he would walk slowly back to his former place. And if he heard the baby cry out, he would race toward the sound, only to be stopped sharply by the cage bars, helpless but at full attention, impotent in his huge masculinity.

Kongo pressed against the cage, his hands clutching the bars. He would watch longingly as the photographers surrounded Lulu's cage and the ice-cream vendors, Georgio and Gus, gave her ices and pretzels. Richie and Eddie actually entered Lulu's cage and played with her. They fed her yogurt and ice cream. They even touched the baby. The crowd oohed and aahed over the baby's every appearance. They waited for hours for her to come out into the sunlight. But only the overflow of people stood in front of Kongo's cage, turning away from him, craning their necks, straining to see the mother and her baby. Even the keepers now merely threw his food to him as they passed hurriedly on their way to Lulu's cage.

Kongo was alone, and he hated it.

A month after Patty's birth the situation came to a head, and once more the apes' lives were changed. Kongo had been watching a girl photographing Lulu and the baby all morning. Intent on her work, she had not seen Kongo watching her. He rolled a bit of leaf between his lips, seemingly inattentive, always alert. Then, finally, suddenly, she was within reach. He flew across the cage, thrust his hand between the bars, and grabbed her arm. The camera flew from her hands, and weak with terror, she started to fall against the cage. Someone watching screamed. The keepers began to run.

In a second it was over. For Kongo let her go. His act had brought the desired results, for suddenly the keepers became aware of his lonely frustration. It never occurred to them to punish him. Within minutes they were bringing him ices and pretzels. They were stroking his hand and scratching his back. He sat docilely by the bars, no longer forgotten.

But the problem was not yet solved, for during those weeks of separation Lulu, too, had grown more nervous. For the most part

28

she cared for her infant, but without any other interest or stimulation in her life, she became bored and frustrated. She wanted the male badly. Always within each other's sight, they would gaze at each other through the bars. Separated by the wall, they sat next to each other for an hour or more at a time, slept there, ate there, just sat there.

Lulu became rougher with her child and would suddenly fling her up onto her back or violently pull her around, swinging her more like a doll than a live thing. Finally, a month after the birth, she threw the baby to the floor and left her there alone. The baby, helpless, lay in the middle of the cage screaming; her mother swung frantically from bar to bar, ignoring her. Once more Fitz called Ken.

"It can't get much worse," Ken said. "This way you'll have to separate Lulu and the baby. The keepers can't stay up all night forever, just to pamper Lulu. She'll just become more nervous. Eventually this has to hurt the baby. You'll have to take her away." He looked at Fitz anxiously. "What can you lose?"

Finally, Fitz made the crucial decision. The family was to be reunited.

This time there were no crowds. Fitz, Ken, and the few familiar keepers were the only ones to see it.

The door between the cages opened. Lulu and Kongo sat in their respective cages without moving, looking through the opened space at each other.

Then Kongo entered Lulu's cage, went to a corner, and sat, pretending to finger some apple slices at his feet, but watching his mate from the corner of his eye.

For twenty minutes they sat apart, evading each other's glances. Then slowly Kongo began to close the gap between them. Little by little he shifted his position until he came to Lulu. They sat side by side. He touched her. She made no objection, and his hand moved down toward the baby on her body. He touched her. Still, Lulu made no move.

Throughout the morning Kongo made slow and gentle advances. Lulu tolerated them only when she could not escape. But she allowed him to stroke the infant's back gently, and she did not put the infant down again. Little by little the animals began to relax,

29

and by that afternoon I saw them resting together, lying on the balcony beneath the windows and touching.

The next day they were playing

Lulu was never careless with the infant again. She handled Patty Cake gently with great concentration and care. And suddenly Kongo was very calm and very adult. He no longer ran or swung wildly around the cage but now walked gingerly, often on the back of his wrists and the sides of his feet, a position not only signifying nonaggressive intent, but one so awkward that it forced him to move slowly and carefully around the cage. In this position he went to look at his infant on Lulu's lap. He peered down at her for minutes at a time, and then, somehow satisfied, he moved away to watch them from a distance or to play. He seldom tried to touch her then.

What was it that had changed him so quickly into this protective, responsible male, watching everything, alert to every sound, dignified with responsibility? How did he really know that this little animal was so fragile and that Lulu, too, must now be treated so gently?

There was not much to say about the infant. She was tiny and nearly hairless. She weighed so little that I wondered if her birth had been premature. She would cling to her mother, not yet separate, although born. As Lulu held her, she would disappear into the darkness of her mother's hair. I would search for some sign of her and find only a little clutching hand, pink-flecked at the fingertips, and perhaps a little ball of a head, supported by her mother's hand. Without the support, her head and limbs would flop backward, and she would be still. She was so still that there was no outward sign that she was, even now, changing and growing.

CHAPTER 4
A Hint of Caring
Second Month

I love routine and the discipline that it can cause. It is really the rhythm of an act or a sequence of acts, a basic primitive reality that is repeated in nature in a myriad of forms. Without some kind of established routine within which to behave, order collapses and chaos is created. This is true for all nature, and just as the apes lived within a routine of their own making, or nature's, my days, too, were made of it. And if someone were to have watched me, they would have found me standing in the same position before the gorillas' cages day after day, making marks on a piece of paper, marks which slowly became drawings.

I soon discovered that the apes and I had a very similar routine. In my own mind, this seemed to draw us closer together. We shared the most basic needs, and our daily lives were organized to provide them.

Both they and I rose at about six thirty. We both socialized with our mates, and, they in their cages, I in my home, ate breakfast. Then the business of the day began. While they continued to forage, and Lulu nursed the baby, I, too, started out to "earn a living." In the wild, the apes would have begun to roam over the land in search for their daily food, while at seven thirty I took the bus from my home in the Bronx. By eight fifteen I had arrived in Manhattan. It was a fifteen-minute walk to the zoo.

31

I passed under the clock each morning, unfastened the chain across the door to the Lion House, and let myself in. As I entered the dimness of the house, the warm animal smells enveloped me, almost as if they were a physical presence, something solid emanating from the rows of cages, a tangle of odors of cats on the right and apes on the left.

"Morning, Luis, Eddie!" I called to the keepers already at work in their own routine of the day.

"Hi, kids," I sang out to the apes in the home cage. And Lulu would glance at me from the wooden balcony below the window. She had already given Patty her first meal of the day and was now waiting to enter the second cage and receive the cereal the adults were given each morning. Often, by this time, the apes would begin to play. In the dim light, in the quiet cool house, I was able to see Kongo and Lulu as they moved around each other and touched each other and came together.

Throughout the morning they alternately ate and played, becoming more active until they were again fed a meal of fruits and vegetables just before noon. Then they rested for an hour, usually sleeping until they woke to forage and play again, although not quite as fervently as they had in the morning. The afternoons were lazier times. At four there was another feeding and then more vigorous play. They lay down to sleep at about nine o'clock. This was their general routine of the day.

Little by little as I stood there, I began to see more than the mere appearance of the animals, and gradually their world opened up to me in those small cages, a world totally new and unexplored.

Time revealed the complexities and intimacies of their lives, intimacies that would be realized only if the animals were watched for successive hours and days. From the complex sorts of activities which the gorillas shared as they lay together, watched each other, and played, I began to see the extraordinarily complicated emotional lives they led in these small, seemingly bare cages.

Startled and unbelieving at first, I watched them, seeing and then recognizing the slow and studied enjoyment of each other and the gentleness with which they touched each other, with the care which we humans tend to believe belongs to us alone.

The play I watched was not that of the usually mislabeled "animal instincts," but was spontaneous, graceful enjoyment of each other; it took many forms, often with no sexual intent, especially then, only a month after the baby's birth.

One would touch the other in passing, and the gesture seemed familiar to me. Lulu would go to Kongo and look into his eyes. Or Kongo might reach out to pull Lulu's leg as she went by. Then it would begin.

Lulu turned to him, holding the baby on her belly under her, and reached out for his head. As they lunged at each other, their mouths opened, and they met in a huge hug, their arms pulling at each other in the same way they did at rest, but now exaggerated with energy and movement. They shuffled back and forth on the floor as they swayed, tugging at each other. And, as the baby clung to her stomach below, Lulu let that hand go, momentarily, to touch Kongo's form above her. The baby clung. Ducking under the bar in the middle of the cage, the animals clutched at each other. Then Lulu ran free, looking back over her shoulder at Kongo, following. He overpowered her figure as he stood above her and enveloped her in his arms. A slow-moving shamble of a dance took place before they both sat, suddenly quiet in the midst of motion. They sat face to face, his hands still around her holding her arms. For a few moments they were still. Lulu's head bent to his shoulder, and then, in sudden activity, she closed her mouth in a mock bite over his arm. And the rush of action began again. In play their mouths were always open, searching and reaching for each other's bodies, mock biting each other's shoulders.

All morning the slow and silent ballet continued, the animals moving constantly from place to place.

And they laughed. It was a laugh without the vocalization of human laughter, but with short expulsions of breath. From the midst of the shambling, playing forms would emerge *Heh, heh, heh, heh, heh, heh.*

They rested together as well, and it was rare that they slept apart. Lulu would go to Kongo, very deliberately put her hand on his head, and then lie down next to him. They lay side by side, face to face, and their hands touched, moving, playing with each other, resting on each other's shoulders. Kongo's hand moved softly over Lulu's blue-black and violet hair, over the baby's head. He patted

Lulu's arm. Lulu pulled gently at his. Or he would follow Lulu until she settled down, usually in the rear of the cage; then he would join her. She would hold the baby with one hand and rest the other on Kongo's shoulder. The two of them would lie there, comfortable in each other's closeness, shifting positions every once in a while, running their hands over the other's body now and again, finally falling asleep.

What made Lulu so casually rest her hand on Kongo's arm? What force brought them together to lie without touching as they sometimes did to look into each other's eyes? Was it possible that their emotional capabilities coincided with their intellectual and physical capacities or specializations? Was it possible that they could care? Of course, I realized that much of what they did was dictated by their anatomical structure. Particular and specialized anatomy permitted particular movement and, therefore, appearance. As Kongo lay on the floor with his legs resting on the wall, his arms folded across his chest, he looked like any teenager on the telephone. Therefore, it was often difficult to separate the appearance of emotional behavior from the physical. However, the fact of their sophisticated emotional behavior was inescapable.

In mid-October I first received a hint of the depth of this emotion. The incident involved not only intellect and emotion, but memory as well—not the memory of a practiced and learned experience, but the memory of an event which had occurred only once before.

Late the previous day one of the guillotine doors between the cages had somehow slipped as Kongo was walking through. It had crashed to the floor, just missing his fingers. He had stood quite still, staring at it, then down at his hands, which had narrowly escaped injury. Lulu had come running to him. She put her hand on his back, and she, too, stared at the door. She had pulled her mate toward the center of the cage, away from the door. After a few moments they had begun to forage as usual. It was a momentary occurrence, and the animals had seemed to forget it within a minute or two.

But the next day, when the door was opened and Kongo was about to go through, Lulu rushed forward. Usually the dominant male first entered a doorway, but this time she pulled him back and

34

shoved him away. It was very unusual. As Kongo stood by and watched, she began to run her fingers around the edges of the door, pushing it up and watching it drop the usual inch or so into place. She looked at it as closely as she could, nervous in her actions. Finally, she stepped to one side, and he went through. As usual, she followed him out submissively.

Each time Kongo went inside and out that morning, he hesitated before that door, pushing it up and letting it drop before he rushed through.

It seemed clear that Lulu had recognized and remembered the real danger of the day before and had been afraid; she had anticipated danger and had acted on it. I could never again believe that these animals' behavior did not include care and concern for each other. And I could not believe that little Patty Cake, who so closely watched her parents act and react to each other, was not affected.

CHAPTER 5
Mother Love
First and Second Months

The birth was rare. And rarer still, Lulu accepted her infant.

It is generally accepted that most female captive gorillas reject their firstborn. Very few mothers care for their infants for even as much as a day. Their lack of concern, sadly, is understandable.

First, most of these mothers are captured early in infancy and are thus deprived of the natural maternal care and attention which seems to be necessary for normal development. They have not even been able to watch other mother gorillas raising their young, and this lack may have cost them the ability to care for their own infants. It seems that motherhood has to be learned.

Too, when they become old enough to interact with other young gorillas, learning their social roles by ordinary play and by watching adult interactions around them, there is seldom another gorilla to watch. Deprived of the opportunity to see normal social interaction, they are often unable to interact at all.

It is also believed that the mother's rejection of the infant occurs as an expression of a neurotic adaptation to captivity. The small empty space allotted to them; the inhibition of their normal needs for finding their own food, building their own nests, playing with twigs, branches, grasses, trees, etc.; the vast amounts of un-

37

fulfilled time; the inability to control their lives to any degree whatsoever—these factors might easily lead to the inability to respond to the needs of a strange helpless creature.

Too, human ignorance of the gorilla's natural behavior with her newborn in some instances may lead to the unnecessary removal from its mother. Very little is known of the normal gorilla mother/infant relationship, and because gorillas resemble humans to such a large degree, it is possible that we expect the gorilla to behave far too much like us. We may object to behavior which deviates from that of a human mother, forgetting the vast differences between the two. Even simple acts such as putting the infant on the mother's head often brings gasps of alarm from the concerned keepers. Certainly, when the mother squats above her day-old child and rubs herself against it, it is sometimes immediately taken away.*

Some of the mother's behavior with her newborn might be perfectly appropriate in her natural state but is either dangerous or inappropriate in captivity simply because of the physical condition of the cage, the hard cement or tile floors, the lack of privacy for the mother and child. There might well be a danger to the infant, but the danger has nothing to do with the mother herself.

There are a few cases in which the mother has killed the newborn outright, but more frequently gorilla mothers put the infants to the breast backward or not at all, hold them awkwardly, not knowing what to do, drag them around the cage by their arms, legs or umbilical cords, allow them to fall, or desert them altogether. On many occasions zoo personnel have seen some sort of apparent mistreatment of the infant shortly after its birth and removed it from its mother to be hand-raised from the earliest stages of its life.

But remarkably, Lulu showed no signs of rejecting her infant. Rather, she seemed somehow to know of the helplessness and the needs of the tiny creature that she carried. And from the moment when Patty slithered out of her body, it was Lulu whose energies

---

* One day soon after Patty's birth, a girl who was visiting from the Philadelphia Zoo told me that the personnel there had seen this behavior performed by another great ape and had immediately removed the infant. The mother was "trying to put it back where it came from" had been their interpretation. I feel that it is a form of masturbation by the mother and not particularly harmful to the infant. In fact, the behavior might satisfy some need of the infant, although this point remains obscure.

38

suddenly surrounded the baby and whose world changed to center continually on her. At the instant of birth Lulu suddenly responded, with total and unquestioning acceptance.

How did it happen? Why did she respond so quickly and so easily, unlike so many other captive gorilla mothers? Was it because Lulu had Kongo? They had known each other all their lives. Could it have made such a difference to her?

Ten to twelve years would pass before Patty would be fully mature. There seem to be three stages of maturity: sexual, at six or seven years; physical, at about nine or ten years; and social, at approximately twelve. According to one of the foremost gorilla experts, George Schaller, the infancy of the gorilla lasts as long as three years before the animal passes on to the juvenile stage, and within that period the mother usually does not give birth again. She must care for her infant.

For those three years of infancy Patty would need her mother's care and attention to accumulate the basic physical skills, knowledge, and experience she would use for the rest of her life. The transfer of that attention to another younger, even more dependent sibling could be harmful to her normal development. In the wild, unable to care for itself, the infant might even die. But for Lulu and Patty that wild did not exist, and their separation was a long way off. For some time to come, Lulu's responsibilities would be to her newborn daughter.

There were to be so many new beginnings in the course of Patty's infancy that I came to feel that her entire life was made of them. For growth takes place so slowly that each new step seemed like some great and important achievement. I reveled in all of them.

The infant was totally helpless. In the beginning she was merely a tiny still form around whom everything happened; she was acted upon. She made no effort, nor indicated in any way that I could see that she even wanted to nurse. On the second day of Patty's life Lulu simply lifted her to her breast. The special little sucking muscles in Patty's cheeks worked in and out; she was nursing. And as she suckled, her tiny fingers soon clutched at her mother's hair.

Of all the anatomical specializations which the gorillas pos-

39

sessed, the most remarkable, at least to me, were their hands. Not only could they grip with the whole hand, but they would move and control each finger as a separate unit. The use of their fingers and toes was so sophisticated that they would pick up tiny bits of breadcrumbs from the floor. It is this ability to manipulate objects which separates the primates from the rest of the animal world, expands the performance of the animals.

Because Lulu could hold the baby in her hands, she was able to care for her in a particular way. From the moment of birth Lulu would lift the baby up to her face to examine her, not only with her eyes, but with her mouth, tongue and fingers as well. At first the baby was so small that she nearly disappeared into her mother's cupped palm. Lulu cleaned her, turning her over and over to lick her every single part, first picking a bit of dirt from her ears, then meticulously from the corner of her baby's eye. Over and over the fingers and tongue were busy on the little body, grooming and caring for every part of her.

Lulu paid a great deal of attention to Patty's genital and anal regions. And soon after Lulu would lick clean and touch her there, the infant would defecate or urinate, as if the mother's actions had stimulated the elimination process.

She put the baby's hands and, less frequently at this time, feet into her mouth; she pressed them gently with her teeth, sucking and licking each finger and toe.

As I watched, I began to wonder if Lulu's constant attention, which often seemed merely to amuse or delight the mother, may not also have provided a benefit to the infant. Could it have served some purpose other than that of a casual pastime for Lulu or the mere cleaning that seemed to be taking place? Could she have been licking off the milk that dripped onto the baby as she nursed? Or was she giving the baby a tactile experience that would prove more important as the baby matured and began to use her own hands in more complex ways? Did it act as a massage to relax muscles stiffened from the compulsive effort of clinging? Had it even stimulated Patty to cling to begin with? Did it sensitize the silky skin on the infant's hands? Reasons and purposes for simple acts raced through my head, exciting explanations for the apparently simple, caring actions taking place before my eyes for so many hours each day.

Even the ways in which Lulu held her baby seemed to serve various purposes and seemed to change as the baby became more capable. Throughout Patty's infancy Lulu seemed to "understand" that she had to lead the baby gradually from one little activity to another, each performed over and over until Patty knew it and was secure. Each skill was gradually added to her previous experience, sometimes replacing a more elementary form with an advanced version. And changes began to take place in Patty's growth, often imperceptible until you learned what to look for.

At first the baby was held so that she faced inward, toward the warmth and security of her mother's body. In that cozy position not only did she continually face her mother's breast, her source of food and energy, her life, but she would also begin to spread her arms out over her mother's belly to clutch her hairs there and learn to cling. Lulu first held the baby against her stomach with the palm of her hand. By the end of the first week Patty's pink-flecked fingers clutched at the thick hair there and were able to hold on. As Patty began to cling more easily, Lulu went to other forms of support. When she first felt Patty make those attempts to cling, she supported her first with her wrist and then with her forearm. Her hands were free for other things. But it was not until Patty was fairly adept at clinging that Lulu draped her over her arm and turned here outward to look at the world.

Lulu carried her baby very carefully as she moved around the cage and often walked on the sides of her feet (a position I had never seen before Patty's birth), as if to be sure she would not make any sudden moves to disturb her infant. She usually held her low on her belly, near her crotch. It was the natural place for the protection of an infant gorilla, kept warm under her mother's belly, safe from the branches and twigs and curious hands of other gorillas, had she lived in the jungle. And if somehow Patty slipped, the danger of her fall would be minimized. It wasn't far from her mother's body to the ground.

As Lulu swung carefully and slowly from place to place around the cage, she folded her legs under her infant's body like a little

41

seat. As she lay at rest, she created a cradle with her body, her knees raised, perhaps one leg crossed over the other, one arm across the baby on her stomach. In the wild where gorillas often make nests ten to twelve feet off the ground, these precautions would have been far more necessary. But even here in the zoo, whether on the floor or on one of the platforms above, it made the baby safe.

As Patty learned to cling, Lulu began to move her to other places on her body, shifting her to the back of her neck for a moment or to her chest or belly or back; within a week after the birth she was gently lifting the baby in a little swinglike motion. As Patty became used to the movement, Lulu began to play with her. She would lie on her back and lift the baby to sway gently in the air for just a moment, her first introduction to space. Patty, nearly completely passive, would dangle there, the little head bobbling helplessly on her tiny gray-naked body. Wide-eyed, she hung there in space above her mother's soft body. Then Lulu would lower her and hug her.

This activity, again, did not serve merely to amuse Lulu, but gradually and, it seemed, miraculously, it began to exercise Patty Cake, introduce her to space, and give her a sensation of activity. Someday it would lead her to move her own body. As Lulu dangled Patty and then began gently to swing her, the little arms and chest muscles were strengthened. It was the first step toward the development of her body; those particular muscles would be needed to support much of the infant's weight as it would try to sit, then crawl, then walk and someday climb. It was logical that these were the muscles that were developed first, but I found it wondrous.

Similarly, as Lulu rhythmically patted the baby's head, seemingly in play, the action would keep Patty alert and begin to strengthen her neck muscles until one day she would be able to left her head by herself.

Soon, in the unrestricted freedom of the air, as Lulu dangled her above, Patty began to squirm. She twisted slightly and moved her head to look at her mother on the floor below her.

As Patty responded, the activity changed slightly, and Lulu no longer merely lifted her up, but began to swing her in the air as well. Then she let go of one hand and a few days later swung her one-handed. Then she went even further and held the baby with one of her feet instead. And one momentous day Lulu took Patty off the floor onto the crossbar and lowered her slowly, carefully. For the first time in her little life, Lulu was not the point of security on which Patty could center.

Each day the infant became a little less rigid, less tense, and soon she moved her legs in space. Her mouth opened with the strain. And one day, weeks after Lulu had first lifted her in play, Lulu graduated the little infant from the crossbar to the top ceiling bars high above the ground. Patty accepted the advanced height as a matter of course and swung easily there.

I never saw Patty afraid. Perhaps it was due to her semiarboreal heritage as well as to Lulu's gradual introduction to space which had made it a natural part of her cage environment.

Now, by the middle of her second month of life, not only was Patty's three-dimensional world established, but her physical activity was increased as well. She was no longer the mere dangling object she had once been but was now tossed, swung, and gently bounced by her mother. As her body began to move, Patty began to laugh. The tiniest *heh, heh, heh,* the first whispered sign of actual enjoyment, came softly from her mouth. By the beginning of her third month, as Lulu held her firmly by a hand or a foot, Patty bounced, tossing and turning in the air, giggling, until she tumbled to her mother's stomach and was held there.

When Patty was nearly two months old, she clasped her feet together for the first time, one little set of pink and gray toes holding onto the other. Only holding at first, but within a few days they were pulling and pushing against each other in a little tug-of-war motion. In the process she forced her stiff legs to move, pumping back and forth.

Now Lulu began to pay more attention to her daughter's feet and legs, stretching them out, unclasping the handlike feet, licking and sucking on them as she had sucked on individual fingers a month earlier. As she dangled Patty above her, she held one of the tiny feet in her mouth and pulled it, exercising the legs as well as the arms now.

Finally, all of Patty's body was in motion.

43

As I write this now, it all seems to have happened so quickly. But it was really a slow, simultaneous unfolding pattern of growth, each phase an unexpected little miracle of its own. I do not mean to imply that Lulu deliberately determined Patty's training. Nature somehow provided the ways of growth and experience through her actions. It all made so much sense.

There was another kind of growth which began to emerge along with Patty's ability to control and use her developing body. It was both the awareness of something outside herself to reach for and the desire to reach for it. It may be true that what I call desire, the inner need for something outside oneself, is necessary to allow the ultimate development of an animal at any point in its life. Without a desire to reach out and accomplish, the process of learning may be slowed. Even the physical development of the animal may be retarded, for without sufficient motivation, it may not have a reason to use its muscles. They do not become firm simply by being. So Patty had to want something, and that something was Lulu.

First of all, it was Lulu's breasts which gave Patty her first reason to extend herself and to use her newly developing physical capabilities. She recognized their existence and their association with her food. One day she no longer simply waited for Lulu to lift her to drink.

Just after Patty was born, Lulu's milk began to flow, and she began fingering her newly heavy breasts and felt the warm milk on her fingers. She tasted it, perhaps imitating her daughter's interest in it or perhaps initiating Patty's interest in it. Since that time she often squeezed it from her nipples in order to lick it off her hand. Now, as Patty lay cradled in her lap, Lulu bent her head over her swollen breasts and squeezed the long, firm nipples. Some of the milk spilled onto her stomach and dripped down onto the wide-eyed infant below. Suddenly the baby stirred! One thin arm waved unsteadily up, struggling into the air. And Lulu lifted her to nurse. When, once more, Lulu put her down, the wavering arm again floated upward. Lulu lifted her again, but only partway, and Patty, her arm still raised and waving, faltered. Then, deliberately, it seemed, she fell forward against her mother's body and clutched her mother's hair. She strained the thin muscles of her body, and she pulled. Somehow, with immense wanting, immense energy, she dragged herself an inch or two up toward her mother's breast. Once more Lulu lifted her, and when she again lowered her daughter onto her lap, Patty again reached up, wanting. She had begun to "do."

The following day, as she and her mother lay resting, she left the protection of the warm, safe belly and began to pull herself haltingly over Lulu's form to the breast and the milk she wanted. Now, as always, she tried first and learned first on her mother's safe, soft body before going out to the more uncertain environment of the cage itself.

Day by day her abilities and her confidence grew, and soon she struggled up to Lulu's breasts and past them to her shoulder. She even traveled slowly and jerkily down her mother's side toward the floor. Even her bent legs helped, pushing against her mother's body as she crept and pulled. Every once in a while Lulu would pluck her up and put her back between her legs. But Patty was immobile for only a moment before, with little control and a great deal of desire, her arms would wave shakily in the air, her little legs would push, her fingers would clutch and her body would slowly move.

Although Patty usually occupied some portion of her mother's body, she was not there constantly. There came a time when Lulu did put her down on the floor. This provided the second reason for Patty's attempts to become more active: her need for the comfort and security, the simple visual reassurance of her mother's presence.

From the time Kongo returned to the "family" Lulu kept Patty close and was careful to touch some part of her daughter's body at all times. She never left her infant, even when Kongo was not in the same cage. She did put Patty on the floor in front of her for a few seconds at a time but then lifted her almost immediately to nurse. In time she put Patty down more and more often and waited until she heard a little whimper before she nestled Patty once more on her belly.

But toward the end of October, as Kongo lay napping in the cage next door, Lulu put Patty down and stood over her for a moment. Then, shifting very slightly, she moved away.

As her mother's touch lifted from her for the first time since

45

Kongo's return, the baby did not cry out as she once had. She lay quietly, her arms outstretched, her eyes gazing upward, her skinny legs floating in the air above her body. Her mother disappeared behind her for a brief moment, and Patty, who could not yet turn or lift her head, was alone. An instant later she was lifted to her mother's breast. It was another important first in her life.

By the last day of October Lulu was leaving Patty for thirty seconds at a time. Lulu usually stayed close by, often sitting just behind Patty's head, watching, until a gentle little whimper was heard. Although Patty was active, almost restless, while held in her mother's arms, she lay quietly whenever Lulu put her on the floor. She followed her mother only with her wide brown eyes.

Her eyes were of utmost importance to her. When she wanted something, Patty looked into Lulu's eyes rather than at the desired object. And Lulu always knew. Simply being able to see her mother eased Patty's anxieties, as long as they were not unusually stressful. And when Lulu heard the sudden cries or small whimpers that Patty used to attract her attention, the mother gorilla turned and came to her. She looked into her daughter's eyes for some sign of her meaning and responded with whatever Patty needed or wanted just then. She always knew. Even in the gorilla the eyes are the mirror of the mind.

Just after Patty was two months old, Lulu moved behind her for a longer time than usual, and the need to see her mother forced Patty to use her slowly developing muscles. When Lulu did not immediately return, instead of a helpless whimper, there was movement on the floor; Patty began to squirm.

Wiggling, waving her arms and legs in uncoordinated motions, she began to pivot slowly on her back until little by little she turned and could see her mother's face. Then she stopped moving and lay quietly content, gazing up at Lulu's eyes. Lulu came to her, touched her foot very gently, and Patty lifted both arms toward her mother.

Now that Patty began to assert herself and to find wonderful possibilities of satisfying her own needs, new activities seemed to occur in spurts. Sometimes activities for which Patty may have been preparing for weeks appeared one after the other in a matter of days or hours. The same day that Lulu first turned away and left her, Patty reached across her body toward Lulu. As she strained to touch her mother, her body began to roll to and fro with the effort. Her fingers stretched, grasping out toward Lulu; as she rolled, each motion brought her closer. Suddenly her hand clutched Lulu's hair. She held on, and with tremendous exertion and strain, she pulled. Little by little she raised her head and then her shoulders off the ground, pulling on her mother's hair. Suddenly she lost her grip and she fell back. It did not matter; she tried again. She had, for the first time, raised herself from that position on her back which was so limiting and which had kept her so helpless.

CHAPTER 6
Letting Go
Third Month

Patty was by now increasingly aware of things beyond herself
and her mother. And she was interested in them. She squirmed and
twisted in Lulu's hands to watch Kongo, who, since his reintroduc-
tion to them, often caught her attention.

Often, as Lulu fussed over and busied herself with her child,
Kongo watched them from a distance. He would lumber over, then
bend over the tiny infant cuddled on Lulu's stomach. He licked her
head or stroked her little back. Sometimes he simply looked down
at her and then softly moved away, a teddy bear of a gorilla, who
looked back over his shoulder to glance again at the little bundle on
Lulu's body, as he settled himself across the cage. Once, when
Lulu had put Patty down and half turned away, Kongo started toward the infant and reached out for her. But seconds before his
hand touched Patty's body, Lulu turned swiftly and picked her up.
Kongo extended the gesture past the spot where the baby lay to a
nearby orange.

He and Lulu played together every morning, Patty very much a
part of the activity, nestled on Lulu's stomach, as the adults
chased or swung around the cage or fondled each other.

But today Kongo seemed determined to make a nuisance of himself. He was restless. He played happily and enthusiastically with

49

Lulu until she turned to take the baby to the back of the cage for the morning "bath" and examination. There Patty squirmed and twisted to watch him as Lulu tried, with great difficulty, to clean her. Patty watched, fascinated, as Kongo gathered garbage from the trough and squeezed it, dripping, through the bars. He slapped it and smeared it and mixed it with his urine. He jumped up and down on the puddle he made of it, watching it splash. But he eventually became bored and headed toward Lulu. She moved away from his insistent touch, and soon he was chasing her joyously around the cage. Finally, she turned and smacked him. As Kongo stood glowering at his unwilling mate, Luis decided to separate them for a while. Kongo trotted willingly into the home cage, and the door closed behind him.

Immediately Lulu put the baby down, and she walked a few steps away. Returning, she swept Patty Cake into her arms and swung up to the balcony. She left the baby there for a moment, swung over to the wall to look in on Kongo, then turned and scooped Patty up. Once more she took her down to the floor level, where again she left her. For the first time since Kongo's return she freed herself from the infant, if only for a few seconds.

For the hour that Kongo was separated from them, she repeatedly put the baby down, moved off, then returned to mother her. When Luis once more let the male into the cage, Lulu swung Patty up onto her back and minced away with her usual small delicate steps.

There were some possible explanations for Lulu's apparently possessive attitude toward Patty. It has been observed in the wild that gorilla mothers do not normally allow other adults, especially males, to handle very young infants, but only to come, as Kongo did, to peer at the infants on the mothers' bodies and to stroke them. Perhaps, particularly because the dominant males were in the most dangerous and responsible position, except at rest, it would be hazardous for the troop, as well as the individual infant, to become attached to them. The adult males seem not to have developed the tremendous desire, as Kongo did, to handle the babies, perhaps because they are preoccupied with their foraging and their own responsibilities and activities.

There is a possibility that Lulu exaggerated her natural protective possessiveness as a result of Kongo's separation from them for the first month of Patty's life. Kongo's interest in Patty may have been exaggerated because of the limitations imposed on his movements, interests, and activities owing to his living in these cages.

The following day, when Kongo was again deliberately separated, Lulu put the baby down and took a solitary swing around the cage. She returned almost immediately and nursed her child. A new pattern was established. From now on Kongo was separated from them for an hour or so each day, during which time Lulu no longer felt the need to keep Patty continually attached to her body. As the days passed, although her eyes were constantly on the infant, she left her more frequently. She came back to check on Patty every few seconds, as if to reassure the little gorilla that she was still there and, it seemed, to reassure herself of her baby's safety.

At first, each time she returned, she she nursed Patty Cake, examined her, and held her closely against her stomach. The nursing seemed to be used for encouragement and reward as well as for sustenance. As Patty became used to being left alone, she no longer whimpered after Lulu, perhaps knowing that her mother would return, for she always did. The intervals between Lulu's departure and return grew longer, up to a minute or two. She no longer nursed her infant each time she returned, but soon she merely picked her up to hold her close. Then that, too, disappeared, and Lulu just touched her lightly before she moved off. For Patty Cake that seemed to be enough.

It was another positive step in Patty's short life, for it forced her to follow her mother with her eyes. Her need for Lulu provided her with new incentive, and she squirmed and twisted on the floor, straining to see her. And as Lulu walked around her, climbed the pole, or went to the balcony by the window, Patty began to try to raise her head.

Her eyes were of utmost importance to her. When she wanted something, Patty looked into Lulu's eyes and not at the object (at the breast, for example, when she wanted to nurse). And Lulu always knew. Simply being able to see her mother was enough security for Patty to ease whatever anxieties, as long as they were not unusually stressful, that she might have. And when Lulu heard the sudden cries or small whimpers that Patty used to attract her attention, Lulu would turn and come to her. She would look into her daughter's eyes for some sign of her meaning, and respond with whatever Patty needed or wanted just then. She always knew. Even in the gorilla, the eyes are the mirror of the mind.

One day in mid-November Patty unexpectedly turned over onto her side. She was very quiet as Lulu swung around the cage. But as Lulu started back toward Patty, the infant began wiggling wildly on the floor, perhaps in anticipation. Her arms and legs struggled in the air. The movement became a rhythmic action that rolled her body from side to side. Without warning she rolled over, then back again. Then she grasped the hair on her mother's leg and tried to pull herself over. But she did not have enough strength and fell back. Lulu picked her up and nursed.

The next time Patty's efforts were rewarded by success.

Consistently she began to roll over onto her side by clutching her mother's hair with her far arm and pulling herself over. Using the same motion, she soon turned without holding onto anything at all.

Also in mid-November Lulu took Patty up to the balcony and left her there. She swung down to the crossbar and sat at the front of the cage, where she glanced back up at the infant and then looked away.

But Patty, with her newfound energy and ability, was wiggling her body and slowly moving across the board, six feet above the hard cement. Bit by bit she moved slowly toward the edge.

Although I usually did not allow myself to interfere with the animals, this time I could not restrain the cry of warning to Lulu.

"Lulu, watch the baby!" I called out, pointing toward Patty. She was wiggling closer and closer to the edge. Miraculously Lulu turned. She looked up toward the shelf, and in one swift movement she swung up to the balcony and scooped up her daughter.

Did Lulu really understand my words or my gesture? I had not expected a response, for my warning had been totally involuntary. But Lulu did respond instantaneously and correctly. The following day the incident was repeated in exactly the same way. Again at my cry Lulu took Patty from the edge of the shelf and away from danger.

The shelf was removed the same day as soon as Fitz realized that

52

the combination of its height and Patty's newfound mobility could lead to disastrous results. What might have been a harmless fall in the soft cushion of the jungle could have led to serious injury or death in the zoo.

On the same day, too, another important change in Patty's life took place. Lulu put Patty facedown on the floor for the first time. Patty had tried out this new position on Lulu's own body, as usual, and had seemed comfortable enough, reaching curiously out of her mother's lap toward the floor. But now she looked bewildered, sprawled out over the floor. Her arms were outstretched, her legs bent and trapped under her. She was unable to move.

Then Lulu turned Patty over onto her back, and suddenly her arms and legs were free to wiggle and wave above her . I realized then that I had never before seen Patty on her stomach. Why? There had to be a reason. In nature there is always a reason.

Some possible explanations came to me. It was only recently that Patty could lift her head. If she had been put on her belly, Patty might have, in her struggles to see Lulu, injured her face or strained her neck muscles. If she had been forced into a motionless position, unable to move those sticks of arms and legs caught under her, her limbs might not have developed properly. Too, not being able to lift her head, she would have forever faced the cement, closed off from the world she was gradually entering. She would have been isolated and, in her isolation, frustrated, perhaps unable even to develop an interest in living. But laid on her back, she had been part of the world which surrounded her. She had wiggled and waved her limbs, exercising them. And she had become interested in, then fascinated by, them. Now she played with her fingers and toes.

Another possible reason Patty had not lain prone before involved cleanliness. For even though Lulu was careful to clear a little spot on the ground wherever she put Patty down, the infant might have ingested the foul material found on the floor of even the best-cared-for animal cage or the bits of earth and detritus found on the jungle floor. She might even have suffocated. Until now Patty had had no need to face the floor. For she was not yet eating solid foods. Nor had she begun to crawl.

But that, too, was changing.

Now that Lulu had put Patty onto her belly the infant was no longer content to lie helplessly, even in that strained and motionless position. She began to struggle to become an active part of her surroundings.

As Lulu sat directly in front of her, watching, Patty weakly lifted her head. It bobbed up and down shakily as she glued her eyes to her mother's face. In another moment Lulu picked up the infant and began to nurse her.

Soon Patty tried to raise herself slightly off the ground. But she could barely lift her head, and her feet were still trapped, clasped like hands, under her. She strained against the floor, but each movement of one leg pulled the other in the opposite direction until she sank back into an awkward lopsided position. She would have to unclasp her feet before she would really be free to move.

Over the next few days Patty began to lift her head and the top of her body higher and higher off the ground to follow her mother with her eyes. Weakly she moved her head from side to side as she discovered and explored the space around her. Her growing need to see things determined her activity. By the end of November she could "sit" for minutes at a time, leaning on her forearms and her knees.

She was growing in other ways as well. Gradually she was able to control the movements of her arms. She no longer lay there, merely waving them around in jerking, undirected movements, but reached for and played with her toes. She deliberately put both arms around Lulu and held on. Both arms worked effectively together for the first time. One day, when Lulu suddenly awakened her from a sound sleep, she struggled against her mother, her arms moving back and forth in a rapid punching motion. They did not move together as a unit as they always had before, but individually. It was a motion and a rhythm which would be important when she began to crawl.

She was beginning to use her hands. As she lay on her back, she played with her fingers, watching them, fascinated. She began to explore her mother, poking her fingers into Lulu's mouth and ears as Lulu had done to her.

By the end of November her arms were strong enough to support her body. And rocking her body back and forth, she began to raise herself up from the ground, supported by the palms of her hands. Suddenly, in one big round movement, she swayed and rolled over onto her back. Lulu turned her over, and again Patty tried to "stand" and to crawl. Her legs jerking back and forth under her, her head held high above the ground, she fell with the exertion and again rolled over onto her back. Again Lulu turned her over. As Lulu nudged her from behind, Patty leaned on the palms of her hands and shakily lifted her body an inch or so from the floor. With her feet still clasped and her nose nearly touching the cement, she pushed her legs jerkily back and forth in her first crawling movement. Her body lurched forward a half inch.

By that afternoon Patty had tentatively unclasped her feet, only the thumb of one foot still holding onto the other. Then that, too, disappeared. As she lay on her back, her feet unclasped themselves, and she lowered one paddle foot to the ground. The other waved in the air above her; she tried to catch it with her hands.

The next day she stood a little higher on her hands and knees, rocking back and forth on bent, unsteady legs.

And the next day, as Lulu held her closely, Patty's feet suddenly unclasped themselves and pushed against her mother's stomach. Now it was not only her straining arms and grasping hands that she used to pull herself upward, but her legs and feet as well. It was as if she were climbing vertically. She stretched, and her back arched with the effort.

At about the same time Patty began to eat solid food. No gorilla mother would purposely feed her baby. The babies learned what to eat, as they seemed to learn so many other things, by observation and imitation. One day there was a hint that it was coming. A piece of banana had fallen onto her mouth from Lulu's, and Patty's tongue had flicked over it, tasting it. Now that she was put down facing the floor, everything there was available to her. By chance Patty leaned her head on a pile of lukewarm oatmeal and now put out her tongue in an exploratory way. She had seen Lulu and Kongo eat that many times. Tentatively she tasted it, then opened her mouth over it. As Lulu picked her up and carried her away, I could see the white cereal pressed messily over her muzzle and her tongue moving some of it around in her mouth.

Her control, reflexes, and physical abilities were beginning to work together. At the end of November Patty saved herself from slipping off her mother's back as Lulu walked quickly around the cage. She grabbed Lulu's leg as she tumbled toward the ground and held on until Lulu put her back up onto her shoulder. Patty was now able to control her body as a single unit.

She was not only carried along under Lulu's belly, fully protected, but often rode astride her mother's back now as well. The world was further opened to her as she peered over Lulu's shoulder to see what she was traveling through. As soon as Lulu began to carry her this way, she automatically ducked under the doors and the crossbars that crossed the cage. Never once did the baby's head strike a bar, even as Lulu ran full speed in pursuit of Kongo. As Patty's strength grew and she was able to lift her head, she began to watch where they were going, lying with her head resting on her mother's shoulder only when traveling at a high speed, at rest, or ducking under a bar. Otherwise, she lifted her head and peered out at everything, curious and completely alert. It was a significant change.

When Lulu played with her, she held the baby by her hand and let her tumble down her arm to swing and bounce in midair. Then Lulu pulled her up to the crossbar again. As she dangled, Patty laughed and squirmed and twisted herself around. She tilted her head back to look up at her mother, who squatted on the crossbar above. She was swung, twirled, and tossed, somersaulted in her mother's hands. Every part of her body was in use.

She seldom protested. She showed no fear even when, for the first time, Lulu ecstatically whirled around the center pole and, as she held the baby's arm, Patty whirled around and around with her mother, her body nearly horizontal in space. She was giggling when they finally slowed to a stop.

But there was a day when Patty's scream cut through the house and shocked us all. It happened on November 9. My father had come to see Patty for the first time. I had been telling him how remarkable the gorillas were and how extraordinary Patty's progress seemed. At noon, when he arrived, there were thirty or forty people in front of the small cage, and we stood on tiptoe at the rear of the crowd, straining to see. Lulu was playing with Patty on the crossbar, lifting her, swinging her, dangling her.

56

Lulu stood up and began to bounce. She put Patty on the bar and licked her stomach. Patty reached up to Lulu and clung to her mother's stomach. Suddenly, without warning, she slipped. She fell from Lulu's body! We heard the scream and the loud smack as her body hit the cement four feet below. Lulu leaped down to her infant. I pushed my way through the horrified crowd toward the cage. My father turned away, suddenly pale.

People were holding their breaths. I yelled for Richie.

By the time I reached the railing Lulu was already holding the baby to her chest. She examined her, turning her over and over in her hands, inspecting every part. Carefully she licked her and kissed her hands and feet. She put the infant to her breast to nurse her, hugging her. At first Patty stared at her mother's face with wide-open round eyes. But in a few minutes her eyes slowly closed, and Patty drifted off to sleep. It was to become her usual reaction to difficult or frightening experiences.

She did not seem to be hurt. I waited for some proof that she was really not injured. In the wild the fall would have been an ordinary happening, part of her growing activity. But there she would have landed on softer ground, covered with vegetation. Here it was a different story, and we watched closely as she slept, hoping that she was all right. Finally, she stirred and began to move around. Lulu hovered over her.

It was early that same afternoon that Patty first turned over from her old position on her back and rolled over onto her side. The fall did not seem to have affected her, and it was an absolute joy to see her insistent activity all that afternoon.

CHAPTER 7
Kongo
Second and Third Months

As Patty clung to Lulu's belly, Kongo watched her from the other side of the cage. He lumbered toward them and sat down, first playing with his mate, petting her as she hugged him. Then, slowly, his huge hand drifted down to the infant on Lulu's lap. Gently he cupped his hand over her little back; then Lulu pushed it away.

He seemed to be constantly aware of the infant. Sometimes he would stroke her, his great, thick finger light on her gray form. Often he bent over merely to look down at the baby as she lay on her mother's lap. Yet, despite his interest, his role with the baby was difficult to define. For although we spoke of Kongo as Patty's father, in the wild it would have been only a biological fact. Gorillas are polygamous. Therefore, Kongo would have functioned not as a familial father to Patty, but solely as a male. He would not have developed any individual parental relationship with her. The tenderness he displayed toward Patty would most likely have been given to all the infants in the group. Perhaps this reflected an innate awareness of the fragility of the small or the young, coupled with the gorilla's natural gentleness.

But this little group in Central Park had so few members and the space it occupied was so small that Kongo's relationship with Patty and his interest in her were exaggerated. It was easy for people who watched them to project their own familial relationships onto the group life of the apes.

In general, Kongo's role as a male was that of dominance. The dominance did not consist, for the most part, of an unpleasant and uncalled-for rule, but was simply a matter of natural function determined by one overwhelming reality. Whereas Lulu was responsible for one tiny helpless being and all her energies were routed toward that central purpose, Kongo functioned for the protection and safety of the group as a whole.

It was imperative that he establish himself as leader and protector and include in his behavior the symbols of that role. Perhaps as a consequence of Kongo's growing maturity or of the baby's birth, his role as the dominant male was emerging.

Kongo regularly took the lead, and Lulu waited submissively behind him to go through the door when it was opened in the morning. At night, when the keepers prepared to close the door, Kongo would lead Lulu inside. They behaved as they would have in the wild, where the dominant male gorilla leads the group from one feeding area to another. The dominant male would first be exposed to dangers; he peers out from the bushes, making sure that it is safe for the troop to cross a clearing; he acts as a crossing guard while the females and the young scamper across, flanked by the other males of the troop.

Lulu often relinquished her place to Kongo and moved to another spot as he came toward her, a part of the primatologist's traditional definition of social dominance. While Kongo usually stayed at the front of the cage to play or observe, Lulu often stayed at the rear to care for or play with Patty. She spent far more time than he did above the ground, balanced on the crossbar. When a stranger entered the house at an unusual time of the day, she would sweep up onto the bars to see who was coming. Kongo almost always remained below.

Kongo was completely alert to any activity in the house, and if the gorillas sensed any threat, any unusual occurrence, it was always he who warily came forward to investigate. If a stranger entered the space between the railing and the cages, which is the keeper's walk, it was Kongo who faced the intruder. Lulu would move back into the farthest corner of the cage, never taking her

60

wide eyes from the possible threat, shielding Patty against her body.

In the wild females can easily climb high into the trees, carrying their infants with them while the males defend them from below. Or the females disappear into the jungle as the males turn to face the enemy. It is imperative that a female be in the safest position and that the mature males expose themselves to danger in order to protect the females and the young of the troop. In the wild, if any particular male is killed, he is expendable, another male can take his place as the dominant figure or the female can join another troop. The mother who has an infant is indispensable; if she dies, the infant dies—and infant mortality is high.

When gorillas are being sought in the wild, the females take to the trees while the males try to defend them from below, where they may be shot. Then the females in the trees may be shot and the babies taken from the ground where they have fallen, sometimes still enfolded in the protecting arms of their dying mothers or piteously clutching the corpse of mothers already dead. Whole troops of gorillas have been casually slaughtered in this way.*

The gorilla's natural defenses against his only real native enemy, the leopard—his fearsome appearance and great strength—are futile against humans. Rather, these defenses have encouraged formation of a myth of ferocity which gives license to kill him. It is the gorilla's appearance, not his temperament, that is fierce, and it is his appearance that he relies on first to defend himself. He generally tries to bluff his way out of a threatening situation and usually must be forced into physical contact with the enemy. Even in the wild most of his charges stop short of the mark, the object being to frighten, not to kill, the enemy. The frightening aspect of his monstrous body, often weighing as much as 600 pounds in a mature silverback, is reinforced by a powerful charging display.

Here in Central Park the intrusion of some unfamiliar person into Kongo's area of the keeper's walk was enough to provoke one of those displays. When a stranger came, Kongo was instantly alert. His eyes would narrow, growing hard and bright as they followed the object of his fear. He would stare at the person, his huge

*It is now illegal to kill or transport gorillas from their territory.

61

head lowered, his lips pursed, and his jaw thrust upward. It was the first sign of his aggression and a warning to the intruder to retreat. But few people recognize these visual communications, or if they do recognize them, they ignore them, complacent in the knowledge that they are separated from the ape by steel bars.

Kongo, still glaring, would deliberately back away and stand in a motionless display of his fearsome magnificence. He was prepared for the charge. He stood high on his knuckles, stiff-legged, his body rigid and hardened, the muscles of his back tense, his head high; his whole body was alive and taut with purpose. Even the hairy epaulets on his shoulders and the hair on his crest bristled. He seemed to grow even larger, even more massive and imposing.

Slowly he rose to full height. With his eyes still riveted on his target, he began to beat his hairless chest softly with cupped hands, the sound hollow. It grew and grew and rolled like thunder as the rhythm quickened and became stronger. At its peak the sound broke. And he moved. Still erect, he charged, slamming the tire violently as he rushed past, dropping down onto all fours. He charged explosively until he was stopped short by the cage bars. There was a sudden jerk of his head, his jaw thrust forward, a loud explosion of breath—*HAH!*—and a futile gesture as if he were hurling grass, sticks, or dirt against his enemy. Finally, he smashed his knuckles against the solid tile wall. And the whole wall shook. The target of this violent power would recoil, stumbling backward. Even the presence of the bars was no longer enough to give him a feeling of security against the stupendous energy of the charging male gorilla.

Chest beating was a very important part of these displays, but the act of chest beating was not limited to times of threat. Kongo beat his chest in different ways according to particular circumstances and moods. Each form was a variation of a basic activity. It was used alone or combined with other actions and sounds. It was done with closed fist (I saw this action only a few times), or open palm. It was performed while he ran or stood still, as a threat in anger, as a release of tensions, as pure exuberance, or perhaps merely as an affirmation of his maleness.

As Kongo displayed his maleness, his dominance, as he strutted around the cage puffing out his chest, as he seemed to grow larger by standing higher on his knuckles, I realized that I had often seen the same outthrust jaw on a young boy about to dare another child to "say that again!" How human these postures seemed to be.

In the year and a half that I watched, I never once saw Lulu take these particular stances, or threaten, or charge. Never once did I see Kongo make threatening gestures of this type toward Lulu. There was something infinitely masculine in them that were his alone.

CHAPTER 8
Male and Female
Second and Third Months

Lulu's femininity seemed to emphasize every gesture of Kongo's as male. Kongo was her protector. She ran to him when she was afraid. Any unfamiliar sudden sound, such as the popping of a balloon, the sound and sight of the blowtorch used by the workmen in a nearby cage, excavation explosions blocks away where the earth was being blasted away for a new subway station, would send Lulu running to Kongo to lean on him. She put her arm around him and turned her head nervously, looking for the source of her fear. Sometimes the deafening pandemonium that often reigned in the Lion House between the hours of eleven and two sent her, in distress, to Kongo. She leaned against him for some kind of solace from the din. By November it had become so noisy and chaotic that Fitz had the bars of the #2 cage covered with boards so that the gorillas could retreat there if they wanted privacy or needed relief from the havoc. (I often wondered if the gorillas got headaches from the racket as I often did by 12 or 1 o'clock.)

There were particular individuals who frightened her for no reason that I could see. One young keeper, for example, had never mistreated her, having had little or no contact with her, but she was terrified of him. Of course, when Fitz learned of her extraordinary reactions to this man, he ordered him not to go into the Lion House.

Even before the young man came into view, she knew that he was there. At the sound of the chain clinking against the door she tensed and then walked toward the front of the cage, where she stood waiting. When she saw or heard him, she backed away over to Kongo and stood erect against him, her eyes focused on the man, her hand on Kongo's arm. She watched anxiously as the man crossed in front of the cage and Kongo followed him. She began to pace, slowly at first, then more and more nervously, faster and faster until the pace became a lope. Then she began to run wildly across the cage. She swung up to the bars, flying through the air, hurtling herself from the bars to the floor, up to the platform and across the top of the cage to the other side, dripping diarrheal discharges. The powerful and acrid stench filled the house. At these times, at the height of her fear, she even seemed unaware of Kongo when he came to her. Her mind was filled solely with her fearful consciousness of this man. As she ran around the cage, Kongo followed, futilely trying to reassure her. Finally, she swung around and clutched him before bolting again. Never still, hooting and barking, she raced from place to place.

The man idly lounged against the railing in front of her. She threw water at him and put the baby on the floor, only to grab her again almost immediately. She stepped in her own excrement and then, frenzied, she rolled in the water and in the filth, splashing it toward him, almost flinging the baby down. When the man moved away down the corridor, she followed him as far as she could, barking. Kongo came to her and put his arm around her. She put her arm over his back and then bent to pick up the now-screaming infant.

The man left. Little by little Lulu calmed down, still swinging around the cage to look for him. For the next hour or so she started at every sound and held the baby to her more tightly than usual.

It took Kongo hours to calm her. And for that time they stayed together. It was he who went to her and lay down near her, who touched her as she moved past him, fondling her as she stopped near him and patting her on the neck to give her the reassurance she needed.

Lulu always seemed feminine. She was not only small, but delicate in her bone structure, her musculature, and her movements. Her body swayed as she walked. And her gait consisted of usually small, mincing steps.

Her eyes, set in a narrow eye case, looked like black half-moons. They were inquiring eyes, and her gaze, when steady, was one of curiosity rather than aggression. Her face was framed in a fringe of hair, even to thin whiskers beneath her chin. The hair at the top of her little flat crest brushed back like feathers into a wispy point.

Lulu often moved her head quickly, like a bird. In fact, her whole body moved with lightness, and when she flirted with Kongo, her whole body flirted and provoked him to follow her. Her anatomy determined the way that she moved. And perhaps her youth. But I never could put my finger on what made her seem so feminine. It was not only her bones, muscle structure, or size, but something else indefinable.

I wondered whether if each animal had been caged separately, their sexual and personal differences would have been so apparent. Or would they have become, rather, less sexually differentiated, ultimately appearing less like individuals to those who watched them? If they were not able to interact with each other, what would have remained to show or explain the essential differences between them?

They did react to each other in a far more complicated way than the simple behavior which many people expected of them. These animals were individuals with personalities and moods that changed according to the varying circumstances of daily living.

Because of Kongo's dominant status, when I gave the gorillas a treat, I usually fed him first. Once he ate he was usually amenable to relinquishing his spot at the front of the cage to his mate. She was hovering in back of him, shifting her feet and waiting for her turn. If I offered a lollipop or apple to her first, she came forward but eyed him cautiously and hesitated. A deep growl or grunt and a low stare were enough to cause her to retreat.

One November morning Eddie called Lulu to the front of the cage. She waddled over to him, looking over her shoulder at Kongo. He was concentrating on peeling on orange, and he didn't even glance up. Eddie put an apple into Lulu's outstretched gray palm.

She pulled it through the bars, moved back to the center of the cage, and lifted it to her mouth. By then Kongo was coming toward her. As his hand reached out, she dropped the fruit, and it rolled to the floor at his feet. Motionlessly Lulu watched him as he picked it up and lumbered back to the spot where he had left the half-eaten orange. He sat down and bit into the apple, chewed it, and let the skin dribble from his lips onto his belly.

"Lulu!" Eddie called to her. "Look! I've got another one for you," and he rolled a glistening red apple across the floor. She picked it up and smelled it. But again Kongo was heading toward her. As she was about to bite into the apple, her mouth closed in midair just above it. She sighed a loud audible *huuuh* and tossed it to him. It rolled to his feet. He picked it up and turned away. He sat down and leisurely began to peel it with his teeth. The first apple lay half-eaten a foot or so away. He reached out and grasped it with his foot. At that Lulu lay down and turned over, facing away from him.

Yet there were many times when Kongo watched Lulu devour a handful of figs or an apple or sugarcane and did not interfere. She seemed oblivious to him. He stood directly in front of her, staring intently at her mouth as she rolled the food back and forth. He followed every movement of her hand as it went to her mouth until all the food was gone. She calmly went on eating, completely unperturbed by him, his nose barely inches away from hers. Many times she stood watching him in the same motionless, apparently covetous manner, staring at his slowly chewing mouth, moving away only when the last morsel had disappeared.

Gorillas in the natural state are vegetarians. They forage, roaming their territory from morning to night, eating, stopping at midday to nap, rest and play. In the Central Park Zoo Fitz tried to imitate the natural conditions of foraging as much as possible. Food was deliberately strewn all around the cage several times each day. The animals had to move about to gather it up. At those times the animals played together the least and were most likely to go their separate ways. Kongo usually gathered his food first and sat with it piled between his legs. He might hold an orange in one foot and an apple in the other, claiming possession, I supposed. Lulu often took her food up to the crossbar and squatted there to eat, perched high above the ground. There she would peel a banana and drop the fruit to the floor below. Kongo would lazily reach over for it, while his mate chewed on the peels, which she seemed to prefer.

They moved casually around the cage, eating, until they were full. Then they either fell asleep or began to play.

CHAPTER 9
Toys and Tools
First, Second, and Third Months

Soon after I began my observations, Kongo came across a long strand of golden straw. He squatted down in front of it, picked it up, and examined it, first sniffing at it curiously as usual. He put the tip in his mouth and pressed the point against his tongue. He held it up and looked at it closely, turned it over, and looked at it again. Then, holding it between his thumb and his forefinger, he slowly inserted the end of the straw into a hollow pore of the cement floor where a single pebble was missing. With great concentration he drew it up and sucked on the end of it. Now, fully involved with the process, he lowered it vertically into the tiny hole and drew it up again to suck the tip. There must have been a minute amount of water in the hole, for as the tip became soggy and used, he bit it off; the strand of straw became shorter and shorter.

It was reminiscent of Jane Goodall's description of the patient chimpanzees using long grasses as tools to stick into nest of termites, the insects crawling up the stem until the chimps ate them.

One morning one of the keepers came in to give Kongo his vitamins. As usual, the ape leaned down to receive them, opening his mouth, his lower lip protruding to catch the orange-flavored drops. But the keeper failed to notice that Kongo was standing not on his wrists, as was usually required of him, but on his knuckles. Suddenly his great hand swept up to grab the dropper.

71

He chewed on it and played with it for quite a while. Then he noticed that Lulu had somehow grabbed the spoon which had held her vitamins and was quietly playing with it. He eyed it covetously; this was far better material than a half-crushed eyedropper, the flavor long since chewed away. He headed for Lulu, head lowered, eyes threatening. Promptly she handed it over to him and strolled away.

He sat down with his new toy and looked it over very carefully. He scraped the floor with it, listening to the metallic scrape that it made on the harsh cement. He liked the sound and made it again. He pounded the spoon and rattled it and clacked it against the floor. And as he did, he began to notice that its shape was changing. The long metal handle was bending. He looked at it very closely, turning it around in his hand. He gave it another whack on the floor, and it changed again. He held it up and scrutinized it carefully. Then, deliberately, he bent it further out of shape. The end of the handle touched the bowl. He looked at it again, and straightened it to its original shape.

He did not continue to bend the stem but stopped precisely when it was in its correct form. Had he understood that the form has a use in one position and less of a use in another?

Now, holding the straightened handle, he again began to scrape the floor with it, bending down to look at the scratches it made. He scratched the tire with it and then came to the front of the cage, where he deliberately fitted the edge of the spoon into the slot in a metal bolt in the cage door. He tried to turn it as if it were a screwdriver!

"Hey, Kongo!" Someone called his name.

His head swiveled toward the cry, and he dropped his spoon. It fell out of the cage, beyond his reach.

Was he using that spoon as a tool? Did he know that he could unscrew that bolt? Had he been imitating the workmen whom he had seen using screwdrivers in precisely the same way that he was using the spoon? He had been interrupted and the opportunity would never arise again.* A few days later he again tried to unscrew some

*These interruptions never ceased to frustrate me. He would never again take up the activity that was interrupted. I felt that a rhythmic pattern had been broken.

screws that held the supports to the plexiglass windows before his cage. And he succeeded enough so that the panel which they held wobbled around in the air.

Lulu, too, continually demonstrated the use of materials as tools, for she often dipped lettuce leaves or paper towels (when she could get them) into some water and then, using them like a sponge, squeezed the water out into her mouth. How had she discovered that soaking a paper towel acts as a gathering device for water? I thought that she had been taken from the wild too early to have remembered seeing other gorillas perform this act. But perhaps she had seen people do similar things.

What a contrast the gorillas' activities were to those of the lions and the tigers across the way! The cats paced and lay sleepily in their cages, tumbling with one another every once in a while, growling and snarling at mealtimes, yawning loudly or roaring rhythmically during the day. But the gorillas played not only with each other in tremendous enjoyment, but also with the objects of their environment.

I got the idea that color might have some attraction for the apes. For Lulu never ate the red beets she was given, but played with them instead. She would sit on the floor with a beet in her hand, stretch her arm out absolutely straight, and with a wide swinging motion slide the vegetable across the floor in a perfect arc. It was as if she were a compass. The deep crimson arc glowed red against the dull, drab cement. Was it just the swinging motion, the normal radial articulation of her arms that she so enjoyed as she swung and slid them back and forth across the floor? Did she ever realize that she had changed the appearance of the floor? She also held the beets in her hands, sliding them across the floor as she ran, leaving tracks of crimson beet juice in her wake. I never saw her do that with any other thing. It seemed as if Lulu used those beets the way a small child uses a crayon when he recognizes what a crayon can do and then revels in making his marks.

Birds seemed to hold a fascination for the gorillas. While they rested, Kongo and Lulu would watch birds, seldom bothering to lumber over to look at the little fluttering creatures. The birds, in their turn, seemed easy in the presence of the gorillas. They hopped around the cage, pecking here and there, disregarding the

73

presence of the huge animals. In fact, in November of the following year a starling happened into the house and made his home there, hopping up and down and onto and into the cages, eating the water bugs that abounded and paying little mind to the apes. It even landed on their bodies and then took off again in flight, completely at ease with them. If the bird came to rest on the bars at the top of the cage, Lulu might sometimes climb up after it, to touch its tail. Was it the motion of the birds that fascinated her? Their delicacy? Their sudden flight as they whisked away? Whatever it was, she did not merely accept their presence, but took interest in them.

In midsummer as I sat beneath the trees outside their cages, a parakeet appeared as if from nowhere. Leaf-green with the delicacy of yellow softening his feathers, he fluttered at the bars for a moment, then perched there. Lulu's eyes fastened on the lovely creature. She came slowly toward the little bird and then sat down, Patty nestled in her lap. Lulu did not take her eyes off the bird. She slipped Patty aside and climbed carefully toward it. Slowly, gently, she reached out her middle finger and brushed its lovely tail. She watched as the bird fluttered lightly around the bars and once more landed above her. Slowly she began to climb toward it again; it took off. It flew into the waving branches of the tree above me. I lost sight of it. But Lulu stayed poised at the bars, spellbound, her eyes gazing on the bird until the leaves stirred and it flew away.

Kongo's concentration on any problem was total. And it was typical of him to take advantage of any situation that would give him just a little more interest in life, a little something more to do. He was smart, and the seemingly empty cage became filled with his inventiveness and his energy. He not only took advantage of the keeper's presence to occupy himself in a number of ways, but used the cage environment itself as well: the nuts and bolts which he managed to work free; the bars; the pulley apparatus that opened the doors to the outer cages. He became so adept at manipulating the chain which hung outside his cage and was forbidden to him that the keepers finally had to tie it to the wire that covered the bars at the sides of the cage, where it was out of his reach. This only created a further challenge for the ape, and with great determination and concentration he was often able to work the wire free and then gently hook the chain with tip of his finger, swinging it until he was finally able to draw it into the cage. Whenever the keepers discovered that he again had the chain in his possession, it added even more interest to life. They would try to wheedle it away from him; but teasing, coaxing, bribing, and the rest of the methods at hand hardly ever seemed to work, and he would swing it tantalizingly before their eyes, then pull it inside again.

Whenever he could involve the men in his games, he did so, for then Kongo had their attention for that much longer. It was for that reason that he so often attempted to get the hose.

"Aha! No, you don't," Raoul or Eddie would cry out triumphantly as Kongo made a swipe toward the nozzle and missed. But when he did not miss, the fun began. Hand over hand he would pull the hose into the cage, gathering it up in the far corner, piling it around and around on top of itself the way a sailor does. Four men, at first, stood outside, trying to pull it out after the first call for help. Kongo would let it slacken and then, with a slight twist of a wrist, jerk the hose back into the cage, winding it around the area like a huge snake. Now six men strained, pulling on it, their bodies tensed against the floor. And Kongo sat relaxed, still holding it easily in one hand. He gave a little tug, barely visible, and the men could not contain it, lurching forward or tumbling backward as Kongo played with them. It would take the six men fifteen minutes of tremendous labor, bribes, and unabashed trickery to get the hose away from the gorilla. It was usually the bribes that finally did the trick. It was one game that I never saw Kongo tire of.

Kongo sometimes seemed to pose problems for himself to solve, and in that way he occupied some of the unnaturally vast amount of unfilled time that the cage situation created. As he became more and more interested in whatever was before him, the intense look in his eyes grew. He became immersed in the problem at hand. If people were not available for his amusement, he simply concentrated on something else; almost anything would do.

One day, as he sat on the crossbar, he played with a piece of parsley, twirling it between his fingers. The lacy bit of green looked almost ludicrous in his massive hand. The brilliant green glowed against the dull black-violet of his sausage fingers. He absentmindedly dropped it, and it fluttered into the trough just outside his cage. Kongo dropped to the floor and, ignoring the bunches of parsley within easy reach at his feet, gazed though the bars at the one sprig he could not reach. He put his hand between two narrow bars and, turning his wrist downward, reached for it. He strained toward the bit of green, brilliant and gay against the dull colors of the cage and the trough. With the tips of his two longest fingers he somehow got hold of it, a tiny ruffle of a leaf grasped tenuously. Slowly and cautiously he began to bring it up toward him, the stem precariously held on by a sliver. His fingers came up to the opening between the bars; they were much too thick to pass between them. He tried to turn one finger so that they could slip through, and the sprig of parsley fell to the floor outside the cage, beyond the trough.

He looked after it for a long moment. And only when he realized that it was too far away did he turn back into the cage and reach down to the parsley at his feet.

CHAPTER 10
Fight!
Third Month

Lulu and Kongo seldom fought. They were usually gentle with each other, and aside from the playful teasing and chasing, there were only a few momentary scuffles or scraps, a snarl or a screech. The few times that I saw any kind of violence between them, its frenzy and fury were frightening.

It was inevitably caused by some outside interference in their lives. People unwittingly created a situation which so threatened Kongo's sense of privacy and safety that his usual attitudes of display and threatening charges simply did not suffice. Once a keeper, unfamiliar to the animals, decided to play with Lulu. An already-excited Kongo dragged his mate away from the man and into the center of the cage. Not too long afterward a similar action occurred, and as Kongo charged, Lulu was in the way. As he attacked, he accidentally brushed her shoulder just hard enough to upset her. She rose, screeching!

The animals grappled, shrieking and lunging at each other in circles of wild pursuit across the cage floor. Patty's grip on her mother's hair was loosened by the suddenness of Lulu's attack on Kongo, and she was dragged along the cement floor. In the midst of the melee Lulu struggled to secure Patty to her again. She ran, screaming, after Kongo. They thrashed on the floor, rolling over and over, now one of them on top, then the other. Patty was lost in the whirling bodies. Then Kongo rose above Lulu and loomed over her thrashing body as she struggled up to bite him. One hand groped for a lever to pull herself up with; her mouth was drawn back in a furious grimace. Outside the cage we watched with horror.

Raoul ran for the hose. He turned it full force onto the screaming animals. They fought in the shooting stream. As they stood there, soaked and shining black with dripping water, their frenzy eased. Torn apart by the power of the blast, they glowered at each other. Then, Lulu, with parting screams, backed away. Kongo, dripping, stood erectly, glaring after her until he was lured to the bars by the promise of milk. Slowly he lowered himself and strode away. Lulu sat in the center of the wet floor and bent over her child, examining her. Patty clung to her now, looking up into her face with round, frightened eyes. It was quiet and calm in the cage once again. And we relaxed.

On a day in November there occurred one of those rare fights caused by growing tensions between the animals themselves.

After their hourly separation that day Kongo greeted Lulu with a slap on the back, a normal salutation. They began to play softly.

They sat down, and Kongo leaned into her nipple. He brushed his finger against it for a moment and then sucked for an instant as he had seen Patty do. He looked at the baby and gently pushed his finger into her stomach. Lulu lightly put her hand on his arm to restrain him. They began to play and she put Patty down for a moment until he moved to touch the infant. Over and over again he reached for Patty Cake, to stroke her with a finger, a look of tender concentration on his features. As Lulu stood up, he cupped his hand under the baby for an instant. He followed them as Lulu moved away. Then Lulu turned back to him, and they played together.

They sat and leaned against each other and hugged. Lulu's hands, which were freed of the baby who held onto her, played caressingly over Kongo. Even as they stood, the hugging became rougher. Suddenly Lulu backed away from him, and Kongo began to run after her. He chased her, grabbing and pulling at her. And she turned to him, this time not to play, but to complain. Her mouth stretched back over her teeth as she cried out, *whoooooo!*

78

whooooo! whoooo! That turned into a scream, *yeeee!* He bounded off, then came back and touched her. She pushed him away, and he hugged her, looking down at the baby on her stomach. Both the gorillas' mouths were open. He pounced on her as if to wrap himself around her and then left, running. He swung down toward her from above. He banged the side of the cage with his hand as he came toward her again. And meeting, they sat, and she hugged him. They stood up to play, then sat again face to face.

Once again they were becoming rougher, and she tried to get away. He held her back and suddenly made a grab for Patty Cake. Lulu tore away from him. She bolted into the private boarded cage and he followed her. Instantly he shot out again and flew up onto the balcony. She chased after him, screaming in fury; the baby was jostled as they ran. Lulu reached up to Kongo, grabbed his arm, and pulled. He flipped off the balcony and hit the hard cement flat on his back with a loud thud. He spun up, and screaming, he tore after her. She furiously turned and met him. Both erect and wild with rage, they met and grappled, the baby somewhere between them. They fought. He pulled Lulu around and flung her off. She pounded and bit him, her harsh, frenzied screams filling the Lion House. He backed into a corner away from her and cowered there. He did not strike out at her, and she hit him again and again. He covered his face with his arms to protect himself. Finally, he struck out blindly at her and ran. She chased him, waddling across the cage, her hand holding her baby to her stomach. She was still screaming, her mouth open, drawn back against her white teeth. She stopped just outside the door of the number two cage where Kongo had disappeared. Then she sat and bent down to the baby. In the midst of this bedlam the baby still held fast to her mother.

Lulu held her up to inspect her, and the baby reached out. Lulu gathered her to her breast, and they nuzzled.

But Kongo was injured. I could see him through a crack in the wooden wall. He sat alone in the middle of the cage, stunned, a wound on the bottom of his right foot. He picked up his foot to look at it closely. Gingerly he touched it, then sat back against the wall.

Soon Lulu went to him, and as he saw her come close, he picked up his foot and thrust it up to her face as if to say, "See what you did." She looked at his foot, then hit him one last time. Calmly she turned and left him there, staring after her. He bent to the cut and began to lick it.

Twenty minutes later he poked his head out the door between the two cages. Limping slowly, he came out and bent to pick up an orange nearby.

Poor Kongo, I thought, seeing him avoid Lulu as they moved around the cage. Every once in a while he sat down, picked up his foot, and cautiously touched the cut.

Luis nudged me in the ribs and winked.

"That big Handsome," he said and grinned. "Always, she beats him up."

Eddie nodded. "He's too fresh," he said righteously. "She gives it to him good."

For the most part Kongo and Lulu avoided each other for two days. They did greet each other, touching briefly when, after their daily separation, the door between them was opened. And Kongo made some slight attempt to touch the baby, who clung to her mother's back. But it was a halfhearted effort, and his fingers barely brushed Patty as Lulu carried her by.

For the most part he kept to himself, foraging and tending to his wound. He glanced toward Lulu as every once in a while she tentatively tried to approach him, but his glance only grazed her, and he turned away, a deliberate snub. But within a day he allowed her to watch as he cleaned the wound and then finally permitted her to pick at it herself. He certainly did not seem to hold a grudge, for soon they had naturally and casually resumed their usual relationship, and everything was normal.

The sore on Kongo's foot was a bad one, about an inch and a half long. He could not step on the wound and so walked on the side of his foot, limping badly. And he kept picking at it.

We watched with growing concern.

I gave him a lollipop the morning after the fight, and he sucked on it slowly, as usual, dipping it into some milk that had spilled on the floor of the cage. When he finished the candy, he held the stick between his short thumb and his finger and poked it into the wound, concentrating on working it into the cut. He licked the blood off the stick every once in a while, and after discarding the stick, he sat and lifted the foot up to his mouth to lick the deep red

80

gash. He stuck his tongue into it, then sucked out the blood, cleaning it, keeping it raw and opened. He picked at it with anything he could find: a piece of glass; his fingers; his teeth; a nail that he had worked out of the wall. And within a week the cut was three times its original size, the flesh around it scraped and torn away.

He worked at the cut with little or no sign of pain, but rather, a complete absorption in the process. He stuck his fingers into it and pulled it apart and inspected it time and time again. And Lulu, once more on amiable terms with him, often bent low to watch closely as he licked it, and she, too, reached out to lick it.

Perhaps Kongo's intense attention to his wound was exaggerated by his captivity: with little else to occupy his mind, he focused on this one area.

He continued to walk on the side of his foot, the foot cupped so that the wound would not touch the floor. Eventually, however, it became infected, and by December 6, it was an inch wide and yellow with pus. It would not heal by itself now and required medical attention.

Dr. Edward Garner of the New York Medical College, the veterinarian for the Central Park Zoo,* was called in to treat Kongo's wounded foot. He bribed the gorilla and lured him to the front of the cage. Eddie Rodriguez temptingly held up an orange, Raoul stood by with a quart of milk, and the big male lumbered to the bars. With much coaxing Kongo finally sat down and put his feet against the bars. He looked like a good-natured teddy bear, sitting relaxed, his big belly protruding, letting the keepers play with his hand. They gave him treats, Ed grabbed the big toe and bent down to examine the foot. He had to duck as Kongo's hand shot out at him and a huge finger hooked his sleeve. Every once in a while the hulking ape swished his hand against some urine in a puddle next to him and sprayed the vet with it. The vet leaned closer; Kongo made a sudden movement; Ed's glasses lay cockeyed across his nose. Finally, tetracycline was prescribed, as well as polynysin B-Bacetracin neomycin powder, which was to be sprayed directly into the wound several times a day.

*Theodore Mastroianni, then deputy commissioner of Parks arranged an agreement of Oct. 12, 1973, between the City of New York and the New York Medical College, Flower and Fifth Avenue hospitals, undertook "to provide comprehensive health service for animals in all zoos under its jurisdiction; and to retain Consultant to furnish such services. . . . " This agreement was, so far as I know, unique. It is no longer in operation.

It was more easily prescribed than done, for as soon as Kongo realized that his foot was to be the center of attention, he took full advantage of the situation. It often took ten to fifteen minutes to spray the injury. He prolonged the procedure, coming to the front of the cage at the sight of the aerosol can, looking at it intently, then retreating. He loped to the back of the cage, then turned to watch the keepers as they went for fruit, Life Savers, or a can of milk to hold up at the bars.

"Come on, Handsome, look what I got for you." Luis popped a piece of candy into his own mouth. Kongo stared at it from a distance. "*Mmmm*, is that good! Look, I got another for you." Kongo sat down and reached for an orange. Then he looked up at the man at the bars and slowly rose. He meandered to the front of the cage and put out his hand.

"You sit down, Handsome. *Ha-ha!* No, you don't!" Kongo squatted at the bars, one hand poised behind him as he stuck out' his lower lip for the candy. Planning to make as much of this a game as was possible, he was ready to snatch the candy the moment Luis' hand was within reach. But Luis was more than ready for his mischief. "*Ha-ha*, no, you don't!" Luis laughed, and seeing that it was useless, Kongo obediently sat down in the proper position; his legs out in front of him, his feet flat against the bars, his hands in full view, holding the bars. Richie grabbed the big toe as Raoul or Luis got into position to spray the wound. Kongo watched them carefully, and just as the man pressed the button on the can, Kongo jerked his foot. The spray missed. Then he pulled his big toe out of Richie's grasp and placed his foot just out of reach.

"Come on, give me your foot. You want a orange, give me your foot." Slowly the foot came forward, and once more it rested against the bars. As one man put the fruit into Kongo's hand, the other sprayed. It was accomplished. But then Kongo lifted his foot, smelled it, and licked off the medicine. The men just looked at one another. They started all over again.

It took a great deal of patience and persistence on the part of the keepers, but soon, despite all of Kongo's shenanigans, the foot did heal.

CHAPTER 11
Kongo and Me
First through Sixth Months

We have finally accepted the fact that the gorilla is generally a gentle animal, perhaps one of the gentlest of all animals. He is gentle to the extent that lone males are able to wander in and out of one troop and then another, although they usually remain on the periphery of the band. Gorillas hardly squabble among themselves, much less attack other species. And this was true of Kongo in the cage at Central Park as well. He was gentle with the squirrel that lived in the tree next to his outdoor cages. If it came into his cage, Kongo would slowly stir his huge body and roll over toward it, poking it gently with one finger before, with a flurry of a feathery tail, it leaped away.

He was gentle, too, with the birds that fluttered in and out of the cage, eating the breadcrumbs that had fallen onto the cement floor. He hardly even stirred or looked toward them. And whereas Lulu might every once in a while smash a water bug that came too close to Patty, Kongo never even tried.*

Kongo was gentle with us, too, as people he knew and accepted as part of his "group." He teased us sometimes for his amusement or to get our attention, and he protected us as he protected Lulu.

*Not all gorillas were as disinterested in the little creatures as Kongo, for in the Oklahoma City Zoo, I was told, the mountain gorilla there enjoyed killing rats.

I had never tried to impose myself on the animals. I did not believe in manipulating them to get a response from them. I had merely been there, quietly working just outside their cages, and they had come to me. It was only after they established contact with me that I even attempted to go into the keeper's walk, an area of Kongo's domain.

"Do they know you?" That was the question that I heard most often from the hundreds of people who filed past the cages all day.

About a month and a half after I had started to draw them I caught a cold. Knowing that gorillas are highly susceptible to humans' diseases and not wanting to give my cold to them, I stayed home for about a week. When I returned, Fitz informed me that Kongo had missed me.

"It was the strangest thing," he said. "Kongo kept looking for you."

Every morning at about eight thirty he had come to the bars expectantly and had waited for the clink of the chain against the door. And when I did not appear, he moved away from the corner of the cage where he came to greet me. For an hour or so he would come rushing to the edge of the cage whenever anyone entered the house. And then he would turn away again.

The day I came back he was obviously happy to see me. When I opened the door, he was waiting for me. And when he saw me come into the Lion House, he sat down at the edge of the cage and put his finger through the grating which lay over the bars. He contentedly half closed his eyes, pursed his lips, and pressed them against the grate. He smiled at me. And then he lay down next to the bars, satisfied.

A short time later when Eddie, in a playful mood, pretended to fight with me within Kongo's view, Kongo's acceptance of me was proved. For Eddie had been one of Kongo's keepers since the gorillas arrived at the zoo six years earlier. Kongo, naturally, was very protective of him and would charge anyone who seemed to be the slightest threat. Yet, as Eddie and I pretended to wrestle, Kongo did no more than glance at us, then away, more interested in his orange than in my mock attack on his lifelong friend.

Eddie stepped back and looked at me with surprise. "You really belong here now," he said. "Kongo, he says you're OK."

84

And in November Kongo proved to me that I was considered a member of his troop. A keeper from another part of the zoo, unfamiliar to Kongo, came up behind me and put his hand on my shoulder. Kongo came alert! He threatened him by throwing his head back, opening his eyes wide, and emitting a loud *HAH!* He pounded the bars of the cage with his fist, and the wall of bars rattled.

And the first time my husband came, Kongo charged him, sweeping urine at him until he ducked away and disappeared into the keeper's room. Kongo watched for him, threatening him every time he came into sight, until, eventually, he came to know him and accepted him.

Gorillas normally spend a great deal of time socializing. Kongo was no exception. He loved nothing better than for someone to play with him: slapping his great hands as he stretched them out between the bars, or scratching his broad back, or merely talking to him. When his company left, he would wistfully watch them go. He seemed to be hoping that they would turn around and stay with him, just a bit longer.

Kongo was fascinated by my equipment—the startlingly white paper and the pencils and watercolors. As I began my day, his eyes fastened on my purse. I took out my paraphernalia and set it on the railing before me. And if I were eating my usual jelly doughnut, his stare was like the one he gave Lulu. If he could have come closer, I am sure that I would have found his face an inch away from mine, his eyes glued to my mouth as I chewed. Lulu, too, would come to the bars and put out her long, thin arm to beg. Soon I found myself tearing off pieces of dough, one for each of them. I put one piece into Kongo's mouth, first making sure, for safety's sake, that his hands were on the floor and he was standing awkwardly on his wrists. It was probably an unnecessary precaution. The times when I put the morsel into his hand I felt the short thumb and finger grasp it slowly and gently. His eyes were still on my face. I put the other piece in Lulu's outstretched hand, and we had breakfast together.

Kongo would take a lollipop with his lips. He would then carry it to the center of the cage and, licking it slowly, watch as I gave Lulu hers. She would thrust her hand out between the bars and snatch

85

the candy from me, far less gently than Kongo did. Delicately holding the stick between her forefinger and thumb, she turned it easily, maneuvering it between the bars. Invariably, while Kongo sucked on his candy, making it last, she bit into hers immediately and chewed it up. In a moment it was gone.

After Kongo had finished his treat, he would lie down next to me by the bars. He stared up at me, not with the angry, threatening glare that was meant to frighten, but with a soft, smiling gaze. As he lay on his back, he would often clasp one foot with his hand. Then he would shift and sit up, putting his fingers through the wire fencing at the edge of the cage. Only the very tips of his fingers poked through for me to touch. And he put his mouth on the wire and pursed his lips at me, silently gazing into my eyes. At times his love of and need for attention were sad, as he sat at the bars waiting for someone to come play with him. Dick Berg often came. He would grasp Kongo's fingers, brace himself against the wall, and pull them. Or he would slap at Kongo's huge hands as the ape clasped the bars between them. And Kongo would playfully try to catch Dick's hands between his own. He loved this game.

He hated the separation between himself and Lulu and Patty which had been considered necessary by the keepers. It would have been unrealistic for us to expect him to be anything but unhappy or bored during those hours alone. In the wild there would have been other females, other companions, responsibilities, and distractions to keep him employed, but here there were no others. The actual process of the separation was filled with as much interest and intrigue that Kongo could create, for not only did he not want to be separated, but it was at this time of the day that he was the center of attraction.

Since it was difficult to convince Kongo to enter his cage alone, within a short time the keepers learned that it was best to convince Lulu to separate herself from Kongo first.

At the beginning, as Kongo or Lulu was tempted from one cage to another, Kongo was wary enough not to be caught for a while, and Lulu naturally stayed with him as he moved. But as the game persisted and he became more excited, Kongo would finally rush into one of the cages and, forgetting his plight, would push Lulu out of the way in his eagerness to get the lovely treat on the other side. The door would slam shut behind him, and he was trapped for the day. When the realization struck him, he spun around and saw the closed door. His fist hitting the metal sounded like thunder.

In time Kongo became smarter and devised games that occasionally made the process painfully difficult for the men but provided the ape with entertainment. One morning, as Raoul and Richie tried to bribe Kongo into the third cage with sweets and other favorite goodies, Eddie opened the guillotine door between the cages. Eddie struggled to keep the chain from slipping through his hands and the heavy door from crashing shut. His body strained, and his face was contorted with effort. When the big gorilla walked toward the door, he hesitated. He glanced at Eddie and then sat down, just inside. He casually put one foot over the threshold. He reached over the doorsill to pick up some fruit that Raoul had thrown just within his reach, and once he stood up and stretched his whole body through the door to get one lone grape on the other side. Then he returned to his position at the door, settled down, and carefully peeled his little grape. He glanced up at Richie, then looked away as if he were bored.

Eddie was getting very tired. The sweat poured down his neck, and he grimaced from the exertion. A vein on his neck throbbed. But he was helpless. He had to hold on and keep pulling.

Finally, Lulu saw a piece of candy on the other side of the door and stepped over her mate in order to get to it. Only then, seeing Lulu take his piece of candy, did Kongo forget, give up the game, and leave his post. He rushed in, pushed Lulu back, and was trapped. As soon as Lulu walked into the first cage, Eddie let the chain slide between his hands. The door closed.

Eddie sank back against the railing, panting with exhaustion and rubbing his aching hands and arms. There were bright-red marks where the chain had pressed into his flesh. With shaking hands he wiped his dripping forehead. "That SOB," he muttered to himself. Then he looked at me, and with a painful laugh he repeated, "That SOB." There was a kind of wonder, even admiration, in his voice.

Now that Kongo spent so much of his time alone, he demanded more attention from us than he had previously. When he did not get it, it was sad to see him unoccupied, lying sullenly on the floor, sulking and despondent. I often went to him for a few minutes at least, making certain that he knew that I was sketching him, for then he perked up and came to the front of the cage to watch me. When I went into the keeper's walk to talk to him, he wanted very much to play and put his hands out as he did with the keepers. But I would not accommodate him. It was not that I was afraid that he would intentionally hurt me, but that his great strength and a sudden, quick movement in play would inadvertently hurt me.

One day we found a way to play without touching. It happened by accident, as so many things do. I had left the side of his cage for a moment to get the sack in which I kept my supplies. When Kongo saw the sack, he misunderstood my intent. As I dug into the bag for pencils, he must have thought that I was searching for a lollipop. When I showed him the pencil, he must have thought it a poor substitute, for he threw a piece of bread at me to show his disapproval. I caught it and tossed it back. He put out his hand to catch it.

He looked down at his feet, at the slices of a loaf of bread strewn there, and he deliberately chose a single slice of bread. This time, with complete attention, he slipped his hand between the bars, awkwardly lowered his wrist, and tossed the bread to me. Again I caught it and gently tossed it back, aiming through the bars at his opened hand. He caught it. We were playing catch!

There was another game that we played—a game which seemed to stress just how alike Kongo and I were in our basic structure and behavior. He had come to greet me at the bars, smacking his lips and beating his chest softly. He lay down and looked up at me, smiling; then he shook his head and his arms gently in a movement of familiarity. I imitated him. His eyes changed their expression; they seemed to sharpen and become more alert. He clapped his hands. When, after three claps, he stopped, I did the same. And when I stopped, he repeated it. For the minute or so that we played, he never broke the rhythm, clapping in groups of twos or threes, then waiting for me to imitate him. There was a particular rhythm which we both recognized and could share. It was also something we did with each other, not some trickery or mischief. We were playing together with a mutual and equal involvement.

One day as I stood near Kongo, drawing and having one of our one-sided conversations, someone called me, and I turned to him.

Suddenly we heard a powerful bang against the wall behind me and a loud, explosive *HUH!* I quickly ducked back to look at Kongo. He was standing rigidly at the bars, glowering at me. He thrust out his jaw belligerently. "All right, Kongo," I said, "I'm not going anywhere." And I settled myself in front of him once again. Once more he lay down, and his eyes narrowed into a smile.

CHAPTER 12
Patty Crawls
December 11, 1972; Fourth Month

There is a moment in the learning process when the physical capabilities and mental attitudes of an animal join and move together as one, each dependent on the other, each urging the other on. Suddenly the animal is ready. Suddenly Patty was crawling.

Lulu had left her sitting in the middle of the cage. She had been watching Lulu intently when somehow her feet unclasped themselves from the tight knot of toes that had kept them bound together and useless. One stiff foot pushed forward jerkily, then the other. One long, skinny arm tentatively slid out to support her weight. Then she moved: an uncertain, halting, and determined struggle to reach her mother.

Lulu leaned down to watch as Patty crept toward her, her legs rigid and bent under her. Inch by inch, she lurched forward. When Patty reached her mother, Lulu picked her up and held her close.

That was another beginning.

In the split second when Patty Cake decided to go to her mother, she turned from a world of dependent helplessness toward one of active independence. The world, even the limited world of the cage, was gradually opening up to her now.

All that day Lulu seemed torn between her mate and her child. She would leave Patty to play with Kongo, returning to her daughter, then going again to Kongo. As the adults played together, Patty sat watching; her head shook, unsteadily tilted back on her body. Then her legs would twitch under her, and she would start to move toward them. When she came close, Lulu reached out to take her. Or if Kongo were not nearby, Lulu would back away a step or two, then stop. Patty would take just a few more wobbly steps toward her. Again Lulu would back away. And again Patty was encouraged to go just a bit farther to reach her mother. Then Lulu would gather her up or bend down to touch her head with her lips, to kiss her.

In the following few days, as Patty continued to creep toward her mother, Lulu continually encouraged her, first by merely moving away from her as she did on the first day, later, by actually restraining the baby and interrupting her progress by picking her up, making it slightly more difficult for Patty to carry through her intention. Perhaps this alternation of encouragement and restraint was necessary and integral to her growth. Perhaps it helped make her strong, for always Patty persisted in her efforts, whether crawling or, later, climbing.

Too, Lulu may have sensed that Patty was not yet prepared to do as she pleased, despite her persistence. Lulu's constant use of restraint served as a reminder to Patty of her dependence on her mother. It was a discipline which would train the infant gorilla to stay close to home even when she was able to move about on her own, for she was not yet able to care for herself. At the beginning, even while Lulu encouraged Patty to crawl, she never allowed the infant to wander more than two or three feet away. As Patty grew older and more adept, Lulu slowly extended the distance between them. Patty's independence grew gradually as the physical space between herself and her mother grew.

Patty's curiosity grew naturally as she became used to her new abilities, and her ever-increasing skills stimulated further activities. When Lulu lay down to rest, Patty began her first explorations, feebly creeping under a lifted leg or laboring over the hill of an arm. Soon she began to crawl in circles around her mother, discovering and exploring her periphery. These were her first excursions away from Lulu, the center of her world.

Once she struggled up onto her paddlelike feet and stood as she

92

attempted to reach Lulu's breast. Leaning against her mother and clutching Lulu's hair, she strained, barely managing to keep upright. Then Lulu cupped her hand under Patty's rear and lifted her to nurse.

As Patty's crawling attempts became stronger and surer, the time when she couldn't move at all faded from my memory. Within two or three days of her first attempts at crawling, I began to wonder less at it and merely accepted it, waiting for the next beginning and wondering what it would be.

CHAPTER 13

Father and Daughter
Third & Fourth Months

Lulu still prevented any direct interaction between Kongo and Patty Cake. When Kongo reached for the baby, Lulu pushed him away or interrupted Patty's play by suddenly lifting her off the ground and strolling away with her. Meanwhile, Kongo's attention would seem to be diverted by a piece of fruit nearby, as if that had indeed been his goal all along. But he would look wistfully after them, while Lulu deposited Patty elsewhere. Patty would sit still for only a moment and once more embark on a stubborn little journey toward Lulu.

Late in November, as Kongo walked nearer and nearer to the baby, Lulu did not reach out to snatch her up, and when he saw Patty alone, he stopped. Tentatively he extended his great hand toward the infant; Lulu made no move to prevent it. He picked her up very carefully.

He carried her awkwardly, dangling her by one arm. He held her away from himself as if he were afraid of this tiny gorilla. Lulu calmly followed them. Four or five feet away he softly put the baby down on the ground, and then Lulu slipped in and took her. Kongo lay down, his hands folded across his belly, watching them.

He followed them around the cages for the rest of the day, touching the baby whenever he could: on the head, on her hand, or

95

on her wiggling legs. When the gorillas finally lay down together, he was affectionate and content, stroking the mother and then the infant in turn.

He managed to pick Patty up again on the following day. First, he looked around for Lulu, and then he gently swung the baby up behind him, as if to hide her from her mother. He turned to the infant and squatted in front of her; Lulu came toward them. He tried to hide the baby by placing his body between the baby and her mother, but Lulu reached down under his legs toward Patty. Kongo meekly stepped back and left Patty exposed to view on the floor. And with one movement Lulu expertly swung the baby up onto her back. Again Kongo's eyes followed them; Lulu matter-of-factly walked away.

Two days after Patty began to crawl, the day began as usual. Kongo and Lulu played with each other. Patty had crawled that morning, whimpering after Lulu and begging to be picked up. Kongo, as usual, had made several attempts to reach her before Lulu did but had failed. Now he watched Patty intently as she ate a lump of cereal. She had not yet discovered that her hands could pick up food or that she could separate the sticky mass of oatmeal into parts. She lay nearly on top of it, concentrating on eating, and it was smeared all over her muzzle. She pushed her face deeper and deeper into the pile of cereal.

Lulu and Kongo played behind her. Suddenly Kongo broke away and lunged toward the baby. This time Lulu was too late.

He did not touch her but stood over her, looking down. He did not move away this time when Lulu came but squatted above the infant. Lulu began to move around them; Kongo shifted too, circling the baby to block Lulu's view of Patty Cake. Yet Lulu did not protest. As Kongo squatted above Patty Cake, studying her, Lulu eventually left them alone together and swung up to the crossbar to watch. Still, Kongo did not touch the baby. But when Lulu again came close to them, he stretched out his arm and touched his mate's back in the familiar gesture of reassurance.

Then he turned to Patty Cake. Lightly, gently he touched her arm. He was cautious with her, as if unsure of himself. But this time, when Lulu tried to come near, he gently pushed her away. Lulu hovered over them, seemingly anxious. Suddenly Kongo left the baby and turned to his mate. As they began to play, he led Lulu into the number two cage, away from the infant. The baby, left alone for the first time in her life, saw them disappear from view. She crept after them, whimpering.

Suddenly Kongo reappeared alone. He bounded over to the little gorilla struggling toward him on the cage floor and stood over her as she continued crawling calmly under him. He leaned down, put his lips to her head, then sat down to stroke her.

Lulu followed quickly. She strode over to them and began to move nervously around them, circling them; again Kongo shifted to hide the baby. Quietly he stroked her head; he seemed entranced by touching her. Lulu paced around them, faster and faster, nearly running. She never took her eyes from the baby and attempted little movements toward her child. But they were blocked by Kongo's mere presence. Kongo, completely absorbed in the infant, seemed oblivious to Lulu's more and more frantic movements. Patty, unafraid, examined something on the floor under her nose.

Lulu was in continual movement now, jerking her body as she turned, running back and forth. She raced circles around them and finally defecated in her anxiety to see her infant. Kongo, braver than before, began to explore the little gorilla. He ran his fingers up and down her back and onto her head. As he became braver, he even touched her anus and pulled her little clitoris as he had seen Lulu do so many times before. But he must have pulled too hard, for Patty suddenly screamed.

Kongo recoiled. Confused, he swung his head around to look for Lulu. As she came forward, he pulled her in toward the baby. Quickly Lulu moved in to take charge and examined their daughter. Kongo hovered over them, obviously anxious and concerned. The examination over, Lulu gathered up the infant to nurse her. Kongo appeared relieved. He moved away to lie down on the other side of the cage to watch his family from there. He had had the baby to himself for a few minutes. I wondered if it had been enough to satisfy him.

The day continued. While I observed Patty's progress through the day, Kongo watched her, too.

I saw that a bit more hair had appeared on the baby's back. She played her fingers over the rough texture of the cement floor, mov-

97

ing the tips of her fingers back and forth over the surface. For the first time I saw her turn her body around in order to crawl after her mother when Lulu changed her course across the cage. Later, while Lulu lay by the door, Patty started to climb over her foot, a mountainous obstacle. She struggled up onto it, her arms stretching out, her fingers clutching, her legs straining as she pushed against the cement floor. Then, when she reached the top, she tumbled down the foot, landing next to Lulu's nipple. She had aimed for it all along. She nursed.

Soon,. however, Lulu decided to wander off, and she left Patty lying on the floor and playing by herself.

And there came Kongo!

He was tender with her, squatting next to her, looming over her body, gently pushing her around until she faced him. He traced her ear with his huge finger and softly touched her fingers one by one. The expression on his face was one of tender concentration. He put out his hand as if to pick her up, then withdrew it as if he had changed his mind. Instead, he stood up, towering over her, then bent down and patted her head. His mouth brushed over her hands. When he stood up, one hand lingered on his daughter's hand, and he lifted her very slightly off the ground and softly lowered her again. She did not make a sound. Then he left her to go to Lulu. In a moment he returned and again bent down to Patty. He put his mouth onto her stomach and licked her, and Patty stretched out her arms to him! For the first time she reached up to be held by him. He toyed with her hands for a moment, and then he picked her up. And this time, when she let out a little cry, he did not let Lulu take her. Rather, he carried Patty Cake into the number two cage, where they were out of sight. Feeling like a peeping Tom, I peeked through the spaces between the boards that covered the bars. Kongo was hunched over the baby. He touched her hands and wiggling feet. Lulu seemed undisturbed and wandered toward them every once in a while, then away, then back again.

Kongo picked up the baby and played with her hands, counting her fingers with a sensitive touch, his movements small and delicate. He sat with his legs spread apart, concentrating deeply on the baby, who lay calmly on her belly between them. And to my incredulity, he picked her up by both her hands. Her legs dangled;

her feet clasped each other as usual. He lifted her up and kissed her on her forehead. Then he carried her into the home cage, put her down, and stepped back. Lulu slipped in and took her baby once again.

Kongo did not take the baby again for a long time.

CHAPTER 14
Illness
Fourth Month

Patty was crawling frantically to reach her mother. Each step was an effort. Her stomach barely lifted off the ground, she reached out one unsure hand, pushed back on one paddlelike foot, and tottered forward. Then she sank to the floor, raised her head toward Lulu, lifted her body off the ground, and started off again. She doggedly staggered forward, and with each successive awkward step she picked up the assurance of rhythm. Her mother was waiting, and when Patty finally reached her and looked up into her eyes, Lulu lifted her to nurse.

But progress does not always go smoothly. One December morning Patty suddenly seemed less active. By noon she did not even try to crawl. Her nose dripped, and her eyes were watering. She was obviously ill.

It had been cold and damp, sleet and hail pounding the Lion House all morning. Gusts of wind swept through the broken windowpanes and through the doors each time they were opened.

Saturday and Sunday, too, were bitterly cold, and the winds blew relentlessly through the building. Patty lay on the floor, huddled on her side, and did not move. Lulu picked her up and put her down again, and Patty did not even lift her head to see where she was. When Lulu tried to play with her, she collapsed in a limp heap on her mother's body. Her eyes became dull and gray.

There was no way to give medicine to Patty directly, for Lulu would not allow the keepers to feed her, and Patty showed no interest in eating. They carefully wrapped tetracycline in Lulu's food and fed it to her in hopes that Patty would receive some of it through her mother's milk. But she barely nursed.

Most of the time she lay still. She seemed to be tiny and helpless again. With her inactivity came an impression of feebleness. She lay huddled for hours on the cold, gray cement, her body shaking visibly now and then as she clutched herself. Lulu hovered over her, put her hand out over the inert form, and watched her. When she got no response, she sat down by her infant. She, too, seemed to be waiting. The keepers watched and worried. Should they take her away from Lulu?

Not only was it frightening to see Patty ill, but everyone felt a responsibility toward this fragile member of an endangered species. If Fitz took her away from her mother, it might mean the end of that exquisite and unique relationship. They might never be together again. None of us had ever heard of a reunion between a mother and infant gorilla separated for any length of time. Yet, if her life was in jeopardy, there was simply no choice; she would have to be taken away. Fitz paced in front of the cage, the question continually on his mind. Though most people would have removed Patty at the first sign of illness, it was the last thing he wanted to do. He decided to wait. If Patty got much worse, if a cough developed, he would have to remove her.

On Monday she was a little better. Once, early in the morning, she tried to crawl and then lay on her side. She showed some interest in moving for the first time in more than two days. She brushed the fingers of one hand with the fingers of the other, studying the movement intently. As Lulu picked her up, she now clung to her mother's rich hair. Her eyes seemed darker, shinier than they had the previous day. Lulu seemed to relax.

She carried the baby around the cage, then took her up to the windows. She leaned forward to peer out at the weather, curious as ever. But Patty was still weak. And as Lulu stared out at the pouring rain and the windswept cages, her daughter lost her grip and fell, screaming.

Lulu catapulted to the ground and swept Patty up into her arms.

102

She held her infant against her breasts and bent her head over her, mouthing the top of her head. Patty lay against her, absolutely still. Cautiously Lulu looked her over, and then tenderly she laid her daughter on the floor. Patty was motionless. Lulu squatted by her, looking down at her, touching her gently. At that moment Kongo came over and curiously peered down at Patty. At the sight or the sense of movement above her, Patty screeched. As Kongo recoiled and retreated, bounding away, Lulu pulled her baby from the floor and held her tightly. Kongo again swung down toward them, wanting to see. And Lulu screamed at him. He started and then withdrew quickly into the other cages. Every once in a while his head would appear around the doorway; he would look toward Lulu and the baby, then would vanish again.

The baby lay on the floor, her body drawn into a motionless fetal position. The minutes passed slowly. As Lulu moved away, Patty did not turn her head to follow her mother's form but remained still, staring into space.

The keepers filled the cage with food, trying to interest Patty in something to eat, just wanting to see her move. A bit of cereal fell on her, but there was no movement or recognition at all. Lulu picked her up and put her down again, and the baby's head sank slowly to the ground. The only movement was the blinking of her eyes.

Finally, at 9:40 A.M., she stirred. She lifted her head and then the top part of her body toward Lulu before she sank again to the floor. Then she moved again. It was hard to believe that only an hour had passed since I arrived that morning.

Little by little her body began to move. Soon she sat and then weakly began to crawl, whimpering, to her mother. Lulu sprinted toward her and picked her up.

Lulu made few demands on her that morning, and Patty began slowly to ease back into her daily habits. She crawled a bit, ate a bit, played a bit. But when, at the beginning of the afternoon, she seemed tired, the keepers closed the house for the day. At one o'clock we left her to rest.

The following day she was better. But she tired more quickly than usual and rested a bit more. Kongo, on the other hand, persisted in chasing Lulu in play. She was simply not in the mood for

103

his antics and finally became fed up. She suddenly flew after him, screaming in exasperation. With Patty clasped to her stomach she stood up and waddled angrily toward him, screeching. She backed him up against the wall. He stood cringing before her, his hands covering his face as if to ward off a blow, helpless before her scolding. She lowered herself onto all fours, gave one last shrill, *Yeeeech!,* and turned away. Cautiously he lowered himself to the ground and looked after her retreating slim back. Then he went off by himself and lay down, his head resting on his arm, watching Lulu from a relatively safe distance. She strolled toward him, put her hand on his head, and lay down by him. But Kongo was not content to lie there quietly. He wanted to play. It had been days since he had had a good romp with Lulu. Lulu gave up. She strolled to a far corner of the cage, where she deposited her daughter, and returned to the male. She lunged at him. In a moment they were at it full force, completely wild in free, uninhibited enjoyment. Patty, in her corner, turned to watch their whirling, shuffling bodies as they danced around and around. Then she lost interest and turned to some Wheatena on the floor.

It was the first time in four days that Lulu did not give Patty her full, undivided attention. Patty was well again.

CHAPTER 15
Learning: A Brand-New World
Fourth Month

One day in December Lulu lay resting and Patty crawled around her as usual. Suddenly Patty turned and crawled away. For the first time something other than her mother had caught her interest, and her curiosity proved stronger than her need to be with Lulu. She crawled toward the object, a tile on the wall, as if she had done it a hundred times before, with not even a backward glance toward her mother. She sniffed at the cool pale-yellow tile, then licked it. She looked at Lulu, who lay leisurely watching, two feet away. Lulu made no move toward her. And Patty turned and went back to her.

Within two days Patty had voluntarily moved away from her mother many times, sometimes venturing out farther, sometimes reaching out only a few inches. She was so feeble that these early attempts were difficult for her, but she persisted. A lifetime of exploration was about to begin. Lulu would help her.

She picked up her baby, set her down facing away, and gave a little push. Patty tottered forward, then stopped in confusion and looked up at her mother standing above. "What did you do that for?" was written all over her face. She started to turn. Lulu reached down, and again one finger of her long hand gave Patty's white tufted rear a gentle push. Patty found herself stumbling forward a step. Again she turned her head, perplexed.

105

It took awhile, but Patty soon understood that her mother was encouraging her to move away. Lulu was teaching her to walk, but she was not teaching her to stray. The baby would be allowed to go only a certain distance from her mother, no more than a foot or two. She was learning to follow, not to lead.

Just before Patty was four months old, Lulu began to direct her, as well as to encourage her. She pinched her daughter's little pointed clitoris from behind, and Patty suddenly scrambled ahead, surprised, but obedient. Patty stopped. Lulu pinched. They moved forward. At the beginning of this exercise Patty automatically turned back to Lulu as soon as she made her few faltering steps. But she soon came to understand that she was to continue to move forward, the little pinch was no longer an impetus, but a signal. She soon knew what was expected of her, and they slowly and painstakingly crossed the width of the cage this way. By the end of December there was no part of the cage that Patty had not traveled. I marveled at Lulu's tolerance and persistence; she, who could swing wildly and leap across half the cage in a single bound, patiently inched her plodding daughter around the cages day after day.

Patty was learning her first lessons in obedience. She no longer dictated to her mother; she was too old for that now. As soon as she was able to move about on her own, obedience became necessary; it meant safety, perhaps even survival. Taught as a natural part of her education, as normal a phenomenon as the physical act of crawling, it was totally accepted by her. It was not something that Patty did not enjoy. It made things not more difficult for her, but easier. For she trusted her mother. Lulu had never done anything to create mistrust in her infant. Knowing that Lulu was always there, Patty now began to wander a bit, freely and unconcernedly. When her mother nudged her from behind, Patty turned immediately, ready to obey.

When she ventured away from Lulu, she was lured, primarily, by her discovery that food existed apart from her mother. Her desire for food had led her to struggle up toward her mother's breast. Now the same desire led her away from Lulu.

Everything that she came across went into her mouth. At least, that was Patty's intention. The size of the object had nothing to do

106

with it. She came across an orange that was as big as her head. She was drawn to it as if nothing else in the world existed. Perhaps it was because it was so lovely and bright in her dull-colored world or perhaps because she had seen Lulu eat so many oranges. At any rate, Patty crept up to it. She had not yet discovered that her hands could be used to pick up things, and it never occurred to her to try. She opened her mouth and stretched it over the piece of fruit. It popped out and rolled away a few inches. Startled, she stared after it. She padded up to it and tried again. Again it escaped her, rolling across the floor. She stopped and watched. Never deterred, she was off again.

She chased the orange around the cage for twenty minutes. She crept up on it from behind and attacked. Again and again it got away from her. Soon Patty was giggling, sliding along behind the orange, totally absorbed in the chase. For once she had completely forgotten her parents, who lay together in a corner, lazily watching her. Patty had found her first toy.

It was difficult to realize that everything was brand-new or near-new for Patty Cake. The texture of the cement against her sensitive fingertips, a bit of food, or a puddle were things with which she had never been in contact. She had never before chased an orange. She had never before held a lettuce leaf in an awkward fist or watched water drip between her fingers. But now, as she accidentally clutched something on the floor, she began to use her hands for something other than clinging to her mother. She began purposely to pick things up from the ground. She concentrated intensely, her tongue peeping out between her teeth, as she slowly closed her fingers over a leaf. Then, with great interest, she carefully opened her hand to let it go.

She used her fingers to scrape dried cereal off the bars or to feel the rough texture of the cement floor. Often, as she thrust out her hand to touch some object, she missed. Her depth perception was not yet highly developed. Coordination between hand and eye and visual judgment of relative space would develop as a result of playful, inquisitive exercise. Everything would be repeated again and again until it became part of her; repetition was essential.

Patty's reason for grasping bits of food had nothing to do with hunger. She had not yet learned that she could pick up food and transfer it to her mouth by the use of her hands. It was merely her interest in the act itself that made her go to it time and time again.

In imitation of her parents, she practiced everything as play long before it was ever put to practical use. Each day brought something new and exciting; sometimes we could see the very beginnings of an action that would months later emerge as an essential part of her life.

Toward the middle of January, when Patty was in her fourth month, Lulu began to seat her on the crossbar. Lulu, of course, never let her daughter go; she kept a large hand wrapped around Patty's middle. Patty began to wiggle and squirm. She stretched out one foot and casually touched the bar next to her with her toes. The foot found its way in between the bars; Patty slid it up and down along the bar and felt the metal with her toes. She was fascinated and began to reach out toward the bar with her other foot. It was then that Lulu noticed and drew her back behind the boundaries of the cage.

Offhand it would have seemed of no particular consequence. The bars were there; she was there; it was logical that they would meet. But I was excited! This casual touch of her foot on a metal bar was her introduction to a new world. It would lead her to cling to the bar as she clung to Lulu. Her grasping abilities had strengthened. And this brief, accidental touch would lead her to a second dimension of space: the heights!

She was already putting her hands on the bars, imitating Lulu's gesture of smearing them with cereal, wrapping her hands around the metal, and sliding them up and down. Lulu would come to take Patty away, and as often as not, Patty would turn back as soon as she was able.

She was continually exploring the objects of her cage. There was always something exciting: food, things like a bit of rubber pipe which her father had torn off from a hose the day before, Lulu, and herself.

She held one foot and scratched her belly with the other hand. She kicked her legs, clasping and unclasping them in the air as she lay on her back. And when she lay on her stomach, her arms outstretched, she watched one hand grab a finger with the other hand, pulling it, turning it over; she put both hands together, then sepa-

108

rated them. She was fascinated by herself, a fascination which, with nature's mysterious logic, was developing her sense of coordination between her eye and her hand.

"Monkey is as monkey does!" children would shout in glee.
"Copycat! Copycat!" they sang out in chorus.
Of course, it was true.

As Lulu untied Raoul's shoelaces, Patty sat watching, transfixed. As Lulu stepped back, Patty came forward, reaching toward Raoul's shoes and touching them. She was learning to "beat" her little chest in an awkward, soft brushing motion. Lulu went to the peephole; Patty went to the peephole. Lulu bent down to slurp up some regurgitation; Patty leaned over and stuck out her tongue. Lulu would chew on the chain at the front of the cage, and Patty sat watching, wide-eyed and wondering. She followed her mother in whatever way she could. It was how she learned. No gorilla mother would hand food to her baby or put it in her mouth. The baby would watch which vegetation the mother ate, and as "mama" reached for some, the baby would also reach out and pull it out of the earth.

It seemed that Patty imitated Lulu in her social development as well because Patty developed no fear or any kind of apprehension of her father. No matter how many times Lulu swept the baby away from the male, Patty still made no move to run from him herself. Perhaps it was a result of her watching Lulu and Kongo play with such obvious enjoyment.

One day, at nap time, Patty came across her father sprawled out over the floor. She stopped, glanced at her mother, who was five or six feet away, then began eagerly to crawl toward him. Lulu pounced and pulled her away. When she lay down to rest, Patty took off again, and with the one-track mind which was developing within her, she headed straight for Kongo. When she reached his bulky form, she began to climb.

Kongo jumped! He must have been deep in slumber, for his dazed, startled eyes danced from place to place drunkenly, searching. They finally landed on Patty Cake, crawling matter-of-factly over his outstretched arm. With a bewildered expression on his face he lifted his arm. The baby rose a foot or two into the air as if she were on a seesaw. He lowered her. She kept crawling up. He touched her very cautiously, then turned to Lulu. He held out his arm as if to say, "Here, she's yours. You take her."

She did. Lulu and her baby snuggled up on the floor for their midday nap.

Kongo sat and stared at the baby, who stared at him from the nook of Lulu's arm. He lay down and closed his eyes. But they opened again to gaze on Patty Cake, who had drifted off to sleep on the floor between the two adults, touched only lightly, now, by her mother's hand. Slowly, Kongo, too, began to fall asleep. Fifteen minutes later with his eyes still closed, one of his hands crept back behind him toward Patty Cake. Very gently he touched her. Then, satisfied, he shifted his body contentedly and sighed. He turned away, tucked his forearm under his head and slept.

At the end of the day Luis turned off the lights, and the Lion House grew dark. Outside a gentle rain began to fall, and the mist turned the sky a dusky, dull pink. The house was quiet at that time in the evening, and the three apes lay together on the balcony, silhouetted against the pink sky. There was a little scuffle as Patty Cake wiggled out of her mother's grasp and scrambled toward the mountainous gray shadow of her father. Lulu pulled her back.

CHAPTER 16
Lulu's Girl
Fourth and Fifth Months

I remember one day in January when Lulu lay on the crossbar playing with her daughter. She lifted the baby high above her and brought her down again, nuzzled her, then sat up to lick the baby's face. As I watched, she opened her mouth wide over Patty's head and engulfed it. My heart jumped in horror, then sank. I had recently read of a mother gorilla killing her infant at birth, biting off its head, its arms and legs. And for that long instant I imagined a headless gray body dangling limply from Lulu's hand.

But it did not happen. Instead, Lulu held her mouth open wide over the baby's face and skull. Then she raised her head, and Patty's face reappeared. Her mouth was open in a happy giggle. She nestled against Lulu's breast and reached up for her. Again Lulu's mouth enclosed her skull, and again Patty emerged laughing.*

Almost every move of Patty's was made under the protective eye of her mother. Lulu kept a tight control. She casually uprooted Patty and moved her from place to place as she herself wanted to move. Since Kongo seemed inclined to approach Patty in lieu of other forms of socialization, Patty constantly found herself flying through the air to be deposited elsewhere. Lulu was unconcerned

*I never saw this interaction again.

that Patty was involved in eating or discovering or playing. She wanted only to keep Patty out of Kongo's way.

It did not specifically bother me that Patty was interrupted while she was eating or playing; she could always find another bit of food or another toy. But it disturbed me that she was interrupted so frequently. Her constant inability to complete her activities seemed to be detrimental to her learning patterns in that important rhythms of activities were continually being broken. The interruptions added extra and unnecessary frustration. Patty was becoming apprehensive, waiting for Lulu's ever-descending hand.

Now that Patty was older and obviously straining for more liberty, we thought it might be wise to keep Kongo separated from the females for several hours each day. After a great deal of discussion Fitz and Ed Garner decided that it was worth a try as long as it did not damage the relationship between Kongo and Lulu. With that in mind, they decided to separate the gorillas from 11 A.M. to 3 P.M. each day.

When the door between the cages was lowered, Lulu and the baby were alone. Lulu shook herself for a moment, put the baby down, and, heeding Richie's playful temptations, came to him at the bars. She trotted back to the door again, picked up the baby, and sat down to wait. But nothing happened. When she realized that the door would not be opened for her, she relaxed her hold on Patty and swung up to the crossbar alone.

She came down once or twice to check on her daughter, and Patty, anticipating the usual interference, braced herself to be hauled away. But Lulu only bent down over her or touched her head, then swung away again.

Tentatively Patty began to creep around the cage, covering a wider area than ever before. She would eat or play without stopping in anticipation of the descending hand. Her concentration on her activities seemed to grow stronger. Even her glances toward her mother grew less and less frequent. There was a gradual release of tension between mother and daughter.

The day after the trial five-hour separation, Patty stood up. Lulu had been dangling her from the crossbar. As Lulu held onto her hands from above, Patty's feet touched the ground. A child screamed somewhere in the house, and Lulu suddenly lifted her up

112

again. The next time she lowered Patty, however, there was no scream, no interruption, and she let her hands go. If only for an instant, Patty "stood," before she dropped to the floor with a little plop. She sat, surprised, for a moment; then she picked herself up and wandered off to a lovely slice of apple. A piece of white paper caught her eye, and she crawled over to it. She slid it on the floor in an arc around her, much as Lulu swept the crimson beets across the floor. She went back to the foot of the pole, looked up at her mother squatting on the crossbar above, and tried to climb. She leaned her stomach against the pole and lifted her hands to it. She wrapped her arms around it and strained against the metal, pulling her body upward and pressing her paddlelike feet downward until she was standing against it, shaky and unsure, but standing. She slipped. And began all over again. She wrapped her arms around the pole and pulled. Her legs were bent at the knees, and she wobbled unsteadily; but little by little her body rose from the ground. Once again her need for her mother had been her impetus for another advance forward. And Lulu now came down to her. Patty let go of the pole, turned to her mother, and climbed up onto one of her legs. Then she crawled away. When she turned again, Lulu was once more lying on the pole above her. Without hesitation Patty crawled to the pole. Lulu came down and nursed her.

As usual one thing led logically to another and soon Patty would stand at the foot of the pole and Lulu would simply reach down, take one of her hands and haul her up, lower her and lift her up again. None of this seemed to disturb Patty Cake.

Patty and Lulu could play joyfully without hindrance. Patty giggled and giggled as she tumbled with her mother or climbed over her, and when the little gorilla's attentions were caught by something else, Lulu let her go.

But when Eddie strolled into the keeper's walk to let Kongo back into the cage, Lulu knew what was about to happen. She flew down from the crossbar, flung Patty up onto her back, and strode quickly to the door to wait.

With Kongo's removal a subtle change in the gorillas' routine developed. Lulu and Kongo still played with each other in the morning, yet it seemed that little by little they stayed apart. Even when the family was together, there was less interaction between Kongo and Lulu. Lulu immediately took Patty to the rear of the cage to play or to perform her usual ministrations. And Kongo often came directly to the front of the cage or climbed the crossbar, squatting there alone. Periodically Lulu would join him for a while, but when she had had enough and returned to Patty Cake, he seemed content. He invented wonderful games that took a great deal of time and used all the materials at hand. The game I most enjoyed watching, I think, was the skating rink.

This was a game which called for specific materials and was therefore performed only during the morning hours. It was then that the gorillas received their cereal, and the gloppy mess was one essential ingredient. After eating as much of the oatmeal or farina as he wanted, Kongo would gather it, spreading it over the floor, or sliding it into the corner. The other essential ingredient was liquid. He used whatever was handy, urine or water from the puddles, the residue of the morning hosing down of the cages. He smeared cereal and liquid into a slippery mess; he slapped it, stood on it, and jumped up and down, sending slush spraying out between the bars. He sat back and smeared it with his hands. Then he rose and tentatively stepped onto it. Somehow it was not to his liking, and he went back to work, concentrating on gathering the slop into a smaller area, then spreading it out again. Finally, satisfied, he strolled to the back of the cage, turned, and taking three sudden running steps, slid across the stuff—just like a ten-year-old child sliding down an icy street. Again he trotted to the rear of the cage, and again he rushed forward, sliding a good three or four feet before he was halted by the bars.

Eventually, tired of this particular exercise, he changed his technique. He stood on one foot, using the other to push off, and he spun around and around, nodding his head, flapping his lips, and sometimes flailing his arms. It was a top-heavy, bearlike, free, and awkward pirouette.

He stopped to stomp into the mess again and gather it into a usable form. He trotted to the rear, took a running start, and slid over the slick. Then, perhaps bored with the game, he scooped up a handful and smeared the bars with it. He clapped his hands and watched the stuff spatter.

He had a glorious time!

Lulu and the baby ignored him, for the most part, from behind.

At eleven o'clock, when the crowds started coming into the Lion House, the animals were separated. When, at three in the afternoon, they were reunited, the apes played together, racing and chasing without restraint. After their last meal they would continue; then, gradually, they would calm down for the night.

Patty loved to play. And with the unthinking, unself-conscious gaiety of any infant, she wanted to do what she wanted to do. She would head for Lulu, who lounged at the back of the cage, idling away the time, her arm over her face, her eyes closed in total relaxation. Patty stopped to look at her, then crept up onto her, reaching out an inquisitive hand to poke a finger curiously into her mother's nose. Lulu would twitch and turn away. Patty hesitated for only an instant and poked her finger into Lulu's ear or her mouth. As Lulu's eyes blinked open, Patty would bend down to peer into them. She licked her face.

Lulu gave up. She sat up in disgust, looking helplessly at her eagerly playful daughter. Patty looked up innocently into her face and sat back. Lulu tried again. She lay down with an audible sigh. As soon as she settled herself comfortably, Patty stirred, on the attack for attention. She climbed onto Lulu's head, biting at her face, licking her, brushing the hair on her crest with her palm. Then she pulled, and Lulu sat up with a start, defeated.

Lulu moved away. Patty followed. Lulu lay down again, and Patty scrambled up onto her, giggling. Now it was a game. She was the instigator, the actor, playing, not played with, acting, not acted upon. She had learned from her mother's activity with her. When Lulu finally lifted her, Patty could not keep still in her mother's arms and wiggled out of her lap to fiddle with a piece of bread on the floor.

She also developed a strategy which seemed familiar. I had seen it a thousand times at the playground. If Lulu seemed to ignore her insistent playfulness for any length of time and Patty could not be distracted to something else, she tumbled over and then waited a few seconds. She watched Lulu to see if her mother had noticed. And if Lulu made no move, she sat there and opened her mouth.

115

Out came a piercing scream. It worked everytime, at least at the beginning. Lulu ran to her, picked her up, and, after she had been fully examined for scrapes or scratches, lifted her to nurse.

As Patty's play and exercise became more of a natural part of her life, her physical adeptness grew. Her ability to grasp with her hands was improving daily. She not only held onto her mother, but now clutched the bars at the front of the cage. Each rod was just the right width for her little hand, and she eagerly reached for it. She pulled herself up as she had pulled herself up at the pole. Her mouth opened with the strain, but in no time she was standing! She shifted her feet and actually changed her position as she leaned in against the bars for support. And then, wavering and unsteady, she slowly let go of one of the bars and stretched her arm toward her mother. Lulu, as always, drew Patty up to her breast.

Over and over again Patty went to the bars, and within two days she was able to shuffle a few inches alongside them. She was able to raise her feet. Suddenly she lifted one foot high off the ground. It waved in midair in the general direction of the bar. But she was still uncoordinated, and her depth perception was still ill defined and undecided. Her foot could not find the bar, and it wobbled ineptly halfway there.

She was eager for the bars and would crawl to them repeatedly. She lunged toward them, often missing them entirely and falling to the floor about an inch away. Undaunted, she would reach out, grasp on, and effortfully haul herself up. She stood there as straight as she could, her white-tufted rear poking out behind. Then she would totter an inch or so to her right or her left.

Her grip was strong, and her will tremendous; she struggled to hold onto those bars. Even when Lulu's hand encircled her body and tugged her away, she held on as long as she could until, because she was unable to fight her mother, her fist would unclasp, and she would jerk backward. As soon as Lulu put her down again, she headed for the bars again.

Toward the end of the month, when Lulu grabbed one of Patty's arms to drag her away from the bars, Patty grabbed hold of a bar with her foot, frantically grasping it and holding on entirely with her toes. There was nothing like determination.

116

By the time she was five months old, as Lulu held her on her lap on the crossbar, Patty casually reached out of her mother's lap for the bars. She clasped them and began to pull herself out of Lulu's lap, intent on attaching herself to the bars, perhaps to climb. She had no fear of height, no fear of falling, no fear of attempting the unknown. She knew the bars by then and reached for them with curiosity and ease. She was able to move in many ways, using her body to travel horizontally or vertically, less restricted or confined by her own abilities than ever before. Obviously she was almost ready to climb. It was only a matter of time.

CHAPTER 17
Climbing
Sixth Month

Two days after Patty was five months old she climbed for the first time. Andrée Boillot and I had been standing together at the bars, speculating on the fact that eventually Patty would just have to climb. We had no idea when it would be.

Andrée, who came to the zoo daily just to see Patty Cake, had become a special friend to me.

She and her husband, Marcel, had lost their daughter two years earlier; her death had devastated them.

Once she had quietly told me, "That first day, when I came across the gorilla's cage and I saw that little baby. . . . Ah, Susie." She paused, and her eyes were lit by tears and a tender memory. "Again, once more, there was such a happiness in my life. Do you know? Am I explaining myself?" She peered into my eyes to see if I understood how deeply she had felt the wonder of Patty's existence. "Ah, Susie," she had said, "she is life to me."

Now she sighed. "Do you see how strong that little baby is now? Look how she goes for the bars." She smiled indulgently as Patty wiggled in Lulu's arms, insisting that her mother release her. When Lulu suddenly liberated her, Patty jerked forward and sat still for a moment, regaining her balance. Then she marched determinedly out of her mother's reach, one, two, one, two! She turned, looked back, and then proceeded forward. As usual, she flung herself toward the bars and caught hold of them. For a moment or so she idled there, shifting from one foot to another, toddling along the bars, stopping, then looking at something at her feet. Suddenly I heard Andrée gasp, "Look!"

Slowly Patty's foot lifted off the ground. It waved in the air and then seemed to propel itself toward the bars, grabbing hold of one; her toes wrapped themselves around the bar. Patty's opposite hand reached higher on the bar it was holding. And Patty strained, pulling on it. Suddenly her other foot, too, grabbed a bar. She slid her hand higher up on the bar. She was climbing.

Andrée and I stood grinning happily at each other. We had waited a long time for this moment; it was lovely that we had been able to share it. We watched all the parts of Patty's body working together. It was another totally normal advance, but how special the moment was.

Patty managed to go up a few inches before Lulu's hand descended. Screaming, Patty held on fiercely, but her foot was gooey with oatmeal, so she couldn't keep a firm hold. Down she came, screaming and clicking furiously in protest until Lulu finally let her go. She aimed for the bars, attached herself to them, and climbed.

Patty's movements were exaggerated, just as they had been when she had begun to crawl. She lifted her leg too high, she grabbed the bar with emphasized verve; she moved her hand searchingly before she caught hold. Her body strained with the effort; her mouth opened wide in a grimace; her thin muscles were taut and taxed as she pulled on the bars. Lulu watched her with careful intensity. She came up behind her child, but this time, rather than wrench her from the bars, she sat down behind her. She put out her hand and cupped it behind her daughter's back as if to be ready to catch her should she fall. But she did not touch her. After Patty climbed a half dozen inches or so, then stopped, Lulu decided that was high enough and took her down.

I never ceased to be amazed at Patty's lack of initial fear. Perhaps she was at ease with this new exercise for the same reason she had been at ease with all the others; she had learned first on Lulu's body, as soft and secure a haven as there ever was. Too, she had watched Lulu climb around the cage in the course of her normal day-to-day life. Once she discovered that she was able to climb, she constantly headed for the bars. It was her primary interest in life, and up she went, more easily each time. But Lulu would take no chances, and as soon as Patty had climbed as far as her mother thought safe, she would take her down, wrapping her hand around her child's middle in order to pull her off. Never deterred, Patty would scramble for the bars as soon as she was released.

Very gradually Patty was allowed to climb higher and for longer periods of time. Soon Lulu even gave the baby a little boost at the bottom to start her off, then sat back behind her and watched. Ten days after Patty began to climb up, she suddenly climbed down, stepping one foot to the ground. For the first time, Lulu did not have to take her from the bars; Patty could do it herself. Soon Lulu sat farther and farther away, finally sitting up on the crossbar while she kept her eye on Patty below. Patty's climbing improved. She no longer had to think of which foot was supposed to do what; the act no longer was performed for itself alone, but she aimed at something. That something, of course, was Lulu.

If Lulu sat on the crossbar, it was there that Patty set her sights and determinedly and slowly climbed. Lulu waited for her daughter, and as Patty arrived, she reached out casually and plucked her from the bar. She would not yet allow Patty to go any higher than the crossbar. It was completely predictable every time; when Patty tried to scurry past her, she would be hauled from the bars, clicking and hanging on tightly perhaps, but wrenched away without much regard for her feelings. Often, when she reached that height, she turned and held out her hand to her mother. She knew that she was supposed to go no farther.

Lulu was becoming far more casual about her daughter's climbing. At times she allowed Patty to climb without direct supervision. She might even suddenly lunge at Kongo and play with him for a few minutes while Patty bounced up and down the bars. Within two weeks Patty could climb for seven minutes at a time; she could climb up and then back down, landing on her feet. She cruised along the bars on ground level for a moment, then headed up to the heights once more. She was no longer thinking carefully about climbing and in her enthusiasm her feet would sometimes slip and lose their grip. Her body suddenly dropped and jerked; she held on tightly with her hands alone. Lulu, always on the alert for a sign that help was needed, swung up behind her daughter and took Patty from the bars. One day when Patty suddenly found herself dangling against the bars, Lulu swept up behind her as usual, but rather than making any move to help, she stood quietly, simply watching her daughter.

Patty was moving her legs. She struggled wildly, not sure how to go about helping herself, but trying. One foot caught hold of the bar, then the other.Lulu moved away silently, and Patty continued merrily on her way.

The following day, as Lulu lounged on the crossbar and Patty drew near her, the baby suddenly stopped. She twisted her body, let go of one of the bars, and reached out toward Lulu behind her. Before I knew what had happened, Patty had swung across and hung in midair from the crossbar, her hands hanging onto the bar an inch away from Lulu's feet. She swung gently in space six or seven inches from the bars and tilted her head up to look at her mother's imposing body above her, seemingly more surprised than anyone else. Lulu reached down and scooped her up.

Andrée and I looked at each other disbelievingly. There had been no overt signs that Patty was ready for such a complicated maneuver. We were barely used to the idea that she could climb. Yet she had executed this feat with ease, determined to reach her destination.

Soon it was not only her desire to be with her mother that acted as the impetus for Patty's climbing, but other things in the cages as well. At first the objects were those things she saw her mother using. She eagerly climbed up to the chain on which Lulu so often chewed. She climbed up the bars a step or two, and as her fingers brushed the chain, it flopped against the bars, tinkling against the metal. Her eyes grew round at the sound. It swung wide and clinked as she again let it go. Patty reached out for it and managed to clasp it after three or four awkward attempts. She pulled on it until it was taut, then changed direction and pulled again. She let it fly, and as it jangled against the bars, she giggled and reached for it again. Eventually she would catch it with her toes, lick it with her tongue, try to grasp it with her teeth. And whenever it got away from her, it only meant more fun.

Patty had been longing to climb higher than the enforced limitation of the crossbar; She had often tried to sneak past her mother's watchful eye. But even when Lulu appeared to be snoozing, her eyes tightly closed, she seemed to know that her daughter had hesitated on the bars and was about to try a leap upward and beyond. The hand would invariably descend.

But one day the hand did not descend. As Patty realized that she had not already been stopped, she almost flew up another foot or so, stopped, waited, then leaped up again. Lulu sat up and then stood as Patty kept climbing. She surrounded her child with her body, yet she did not touch her. Patty kept going. She glanced at her mother every now and then but did not stop until she came to the curve in the bars at the very top of the cage. Suddenly she found herself upside down, hanging by all fours like a scrawny four-legged spider on the ceiling. She had never before been in this helpless position and had no idea of what to do. She merely hung. Lulu's hand encircled her body, and Patty put her arms around her mother's neck. They climbed down together. Lulu nursed her, put her down, and Patty trotted toward the bars.

Nothing could keep Patty from climbing for very long. Whenever Lulu permitted, she climbed to the top and then was helpless there. After the first trial runs Lulu no longer followed Patty as closely as she had, and as Patty hung upside down in the curve at the top, she leaned back into a near backbend to look at her mother, watching her below. A little whimper called her mother to her, and Lulu rose to retrieve her from the heights. But Patty was smart, and it was only a few days before she anticipated the curve and stopped before it began. She looked at it; I could see that she was mulling the situation over in her mind. She began to move backward. Very slowly and very carefully Patty backed down the entire long length of the bars. Lulu remained where she was, watching. When Patty reached the bottom, she scampered cheerfully to Lulu and lunged at her, chuckling with glee.

She almost ran up the bars now. And one day, as she was bouncing along, thoroughly enjoying herself, she forgot all about the curve. In her enthusiasm she went too far. Her feet were unable to retain their grip. *She dropped!* Her body jerked hard in space. Somehow her hands clutched the bars, and there she hung, only air between her and the hard cement ten feet below. We watched in terror as she swayed, helpless and tiny, so far above the floor. Her eyes widened in real fear. She screamed! She dangled powerless to move, gripping the bars with all her strength.

When her mother's hand enclosed her body, Patty threw her

arms around Lulu's neck, hugging her fearfully as Lulu brought the baby down to safety and comfort. She whimpered. Lulu held her more closely. She had to unwrap her daughter's arms from her neck. But within moments the fear had dissipated, and in another moment Patty was running off to climb again.

Day by day Patty's climbing improved. At the beginning of March, knowing that Lulu was always there to help, Patty climbed to the ceiling bars, and after one look for reassurance toward Lulu below, she let her feet go deliberately, voluntarily hanging in space for the first time. She began to sway. But it was not the usual, helpless little dangling swing which inevitably led to a cry for help. The swing grew wider. I began to realize that she was actually trying to swing herself back toward the bars. She was trying to control her body in a brand-new way. She strained and pulled to swing higher and higher, trying to carry her body up toward the bars.

She could not do it. One hand slipped and lost its grip. Again she was left hanging by one hand alone; again she was helpless. A little shriek brought Lulu to her aid. But of course, after a minute of comfort, she was off to try again. Once more she lost her grip at the crucial point. But this time she did not call for "mama." This time Patty was determined to regain her hold, and she reached for the bar above her. She stretched, reaching up, and her fingertips touched it. A little farther . . . just a bit more. She caught hold. She was back in position. But it wasn't enough for her. She began to swing. Higher and higher she swung, her toes aimed at the bars. Suddenly one foot took hold! Then the other! Lulu stood ready to help. This time it was not necessary. Patty hung onto the bars with all fours and walked upside down to the curve. She came down all by herself. Lulu sat down and relaxed.

There would still be days when Patty would get herself stuck at the top and Lulu would have to come to the rescue. It would take further practice and strengthening to become consistently sure and safe all the way up there. But each time she scurried up the bars and ventured out she was improving. She moved without fear, without hesitation. She climbed and swung with eager enjoyment, the enjoyment of a healthy, happy, and securely cared-for baby.

CHAPTER 18
There Is a Purpose for Life
First through Sixth Months

Patty was not the only member of the group to benefit from a family relationship. She innocently gave as well as received. For Kongo and Lulu delighted in her presence. Kongo's attentions to Patty seemed primarily to gratify his own curiosity and pleasure. She added interest to his life and perhaps a greater sense of responsibility than he had had before her birth.

For Lulu the essence of her life was changed. Her involvement with Patty was something that went far beyond amusement or enjoyment. Patty reached for Lulu and was content; Lulu reached for Patty and found purpose. Their lives were an unending series of responses to each other, responses that grew or changed as Patty grew older and progressed.

Lulu flourished in it. She seemed to me to become more feminine, engrossed as she was in Patty's well-being. Their sweet and ordinary responses to each other came so naturally that my belief in their ability to have emotions became stronger and stronger. And bubbling inside me, growing larger daily, was the idea that emotion served a very real purpose. It was, I thought excitedly, another specialization which helped ensure the survival of the species, the way in which nature assured the care which each infant of the species needed in order to survive.

The care Lulu lavished on Patty seemed to be born from a fascination with her. But where had this fascination come from? Why did it burst forth with so much intensity and energy upon Patty's birth? Did it emerge partially as a result of the chemical, hormonal changes that took place in Lulu's body?

The effect of Patty's birth reached out beyond the boundaries of her own cage, beyond Kongo and Lulu's lives. It spread into the cage next door where two elderly female gorillas lived.

Joanne and Caroline had arrived at the Central Park Zoo in 1942. They grew old together. At first they had battled fiercely for dominance; Joanne had won; from then on they had lived in a truce. For the most part they ignored each other, sometimes snarling or screeching over food or space. The major part of their lives was spent eating or drifting from one spot in the cage to another, changing position for no particular reason except to move. Their nearly meaningless lives made them inactive.

Joanne, now over forty, was one of the oldest gorillas in captivity. She was enormous; her stomach was distended by the tremendous amount of food that she consumed. As she lumbered heavily across the cage, her belly swayed ponderously under her. She settled herself with difficulty amid the food. She ate mechanically and without gusto, as if someone had turned a key in her back; as the key unwound, her arm reached out and her hand hunted out a bit of food and brought it up to her mouth. Strings of apple skins and slivers of orange peel fell haphazardly onto the shelf of her stomach, a colorful splatter of confetti. She hauled herself onto the balcony, puffing with exertion. She sat, staring out the window, clasped the window bars, and leaned her head against them. Eventually her eyelids drooped, and she snoozed. This was her life.

While Joanne plodded and labored, Caroline paced skittishly. She seemed continually agitated, restlessly moving here and there. As she threw a quick glance over her shoulder toward Joanne, one arm shook with a nervous tic. Her age was beginning to show in her graying whiskers and the thinning hairs at the top of her crest. Her narrow face seemed set in a perpetual scowl.

Joanne acted oblivious to her presence most of the time, and even when Caroline came up to her on the narrow balcony, she refused to budge an inch. When Caroline precariously edged herself

126

around Joanne's stolid bulk, Joanne lifted her upper lip in a semblance of a snarl.

They were living their separate lives in a cage that was eighteen by nine feet. They had been living that way for thirty-two years.

One day in September, at the sound of Patty's cry, the old ladies woke from their dreary tedium, boredom, and indifference. Caroline rushed to the metal wall, hooting softly in excitement. She dashed from one peephole to another in order to see the baby. After all these stagnant years of weary monotony, she came alive.

Hour after hour, day by day, she sat by the wall. She put her eye flush to one of the tiny peepholes that ran along the metal. Her left arm shook often when her excitement or her anxiety grew high. When Patty Cake disappeared from her view, she would shift her body or look for another peephole for a new vantage point. If Lulu left Patty on the floor alone for even a moment, Caroline would grunt her displeasure, and if she heard the baby cry out, she would rush to the wall. Filled with concern and responsibility, she hooted her warnings through the solid barricade. Only when she ate or when Lulu took Patty into the other cages would Caroline leave her post.

Even Joanne showed some interest in the baby. She would lumber over, her grotesque stomach swinging heavily. Upon her approach, Caroline would nervously retreat, and Joanne would take her place, peering stolidly through to the other cage.

Why did they do it? Had they remembered that thirty-odd years ago they had heard an infant cry in the wilderness?

Lulu and Caroline were friends—as much as they could be under the circumstances. They often held hands, each on her own side of the wall. Caroline stretched her arm between the bars and around, until she could catch Lulu's hand, in turn reaching toward her. They sat with hands entwined, an awkward but perhaps comforting position.

With the birth of Patty, Caroline's closeness to Lulu grew. These animals had so little to occupy their time that whatever found its way into the cage was immediately seized upon to be used until it was devoured or destroyed or taken away from them. Yet, as Patty grew older and Caroline's interest in her neighbors was awakened, she began to make gestures of giving. Things that she once had

127

treasured and held to herself, she now gave to Lulu, her need to have some form of sociality suddenly more important than her solitary idle play. One day in February Caroline came across a long piece of straw. She played with her precious find for a while, rolling it back and forth in her hands, examining it as she poked it into a hole in the cement, pricking its point against her palm. Then she stopped. She glanced over at the impenetrable wall. She rose and strolled deliberately over to her favorite peephole at the front of the cage. It was at this hole that she and Lulu often met. Her arm shook; she peered through to the empty cage on the other side. She shifted her position, turned around to watch for Lulu as she came into the cage, then turned again. She fidgeted impatiently. She was obviously waiting for Lulu.

In a few minutes Lulu, with Patty clinging to her back, wandered in. Caroline grew alert, and as Lulu settled down on the other side, she very carefully inserted the thin straw into the hole. It tickled Lulu's ear. Lulu reached back to scratch. Caroline wiggled it again, and again Lulu scratched. She put Patty on the floor in front of her, and the next time she felt the little prick on the back of her neck, she turned. The long, slender straw was wiggling and waving at her temptingly. She looked at it closely for a moment, then grasped it and drew it through. Caroline, on the other side, felt the tug and let go. She watched Lulu play with it. She hooted her pleasure and squirmed to get a better view. Lulu used it like dental floss, wrapping it around her teeth and pulling it back and forth. She pressed it against her foot and then, like Caroline, rolled it between her palms. She dropped it. And Patty, who had been squatting in front of her, watching every movement, picked it up. Caroline came to attention. She stood up and gave little short, sharp hoots of concern. Her arm actually vibrated in her nervousness. She stared intently through the peephole. When Lulu took the straw away from her daughter, she relaxed again. I could almost hear a sigh of relief. She once more settled down to enjoy Lulu's play. Finally, the straw was chewed and broken into a soggy useless string, and Lulu tossed it between the bars and out of the cage. She picked up her daughter and without a backward glance left the cage. Caroline waited by her peephole for a minute or so; then she, too, stood up and walked away.

It was possible, I thought, that I had misinterpreted what Caroline had done. Perhaps she had simply been fiddling with the straw. Perhaps it was all simply a matter of idle play. But soon afterward a similar exchange took place, and there was no longer any question in my mind that it had not been an accident.

This time it was not a mere toy that was involved, but food—and a very special treat at that: hard rolls, crisp and brown on the outside, a cloud of soft fluffy dough on the inside. When Caroline saw Lulu receive her wonderful treat, she began grunting in delight-filled anticipation. Even before Luis reached her cage, she was at the bars, her hand reaching for hers. As he handed the rolls to her, she uttered sounds of pure pleasure.

She took her delicacy to the wall and tore the roll apart. She seemed to relish every morsel that rolled over and over in her mouth as she chewed. Crumbs sprinkled over her stomach; she popped the soft center into her mouth. Suddenly she realized that Lulu had joined her on the other side. She peered through the hole, then turned back to her roll. She plucked dough from the very center of one roll and poked it into the hole. She pushed it through to the other side with one finger. Then, as Lulu saw the dough, Caroline hooted with another kind of pleasure. Lulu slowly worked it out of the niche and popped it into her mouth. Again Caroline tore off a piece and pushed it through. Again Lulu took it. Until all the rolls were gone.

It must have been that Patty's birth freed some long-hidden yearnings inside Caroline, some need for closeness. Even if the baby was not hers, even though she would never hold her, even though she would live separated from her for all her life, she became a part of Patty and Lulu. Life was expanded for her just a little.

CHAPTER 19
Struggles
Fifth Month

I stood in front of the cage struggling to work. Jostled by scores of schoolchildren, I already had a headache. And it was still early in the day. I squinted to see even the most shadowy forms of the gorillas as they moved around. The bars of their cages were painted black, and sometimes the animals seemed to blend into the walls. On top of that, sheets of plexiglass had been placed in front of the cages like windows. They were not there, as many people believed, to protect the people from the animals, but, ironically, to protect the animals from the barrage of candy, lighted cigarettes, peanuts, razor blades, and sharpened paper clips that were continually tossed into their cages despite the sign on the bars that read, DO NOT FEED OR ANNOY THE ANIMALS.

The windows did protect the animals, but they were a nuisance to me. In no time they were scratched and opaque. And the sunlight hit them, dazzling my eyes. I was the only person in the city of New York who prayed for gray, dull, rainy days.

The cement under me was hard, and I shifted back and forth with aching feet, searching through my drawings for one to work on.

"Why don't you work from photographs?" people asked. It was

simply not my way. Whenever I tried it, the life that I loved so much in the animals was lost.

Again I squinted into the depths of the cage. "Wait," I told myself, "just wait until Kongo starts throwing urine all over it." I almost groaned in anticipation of that familiar occurrence. I flicked my pencil against the paper to capture the direction and thrust of some hair on Kongo's crest. A quick, upward stroke of the pencil was needed. Concentration and patience—those were the requisites for work like this.

I felt a tug on my sweater and looked down.

I had been somewhat aware of a shadowy form next to me, watching me steadily in the moving flow of the crowd. He had been there for some minutes. I looked down to see a boy of about ten watching me seriously.

"You shouldn't do that," he stated flatly.

"What?"

"You shouldn't do that." He was very positive.

"Why not?"

"The animals are moving. It's too hard."

I was startled. It had never occurred to me. I looked at him closely. He had some schoolbooks under his arm, and I managed to read the titles of two of them: *Arithmetic Made Easy, Grammar for Fun.*

"Don't you ever try anything that's really hard?" I asked him.

It was his turn to look startled. "No!" He sounded surprised. Before I had a chance to ask him why, he ducked through the crowd and disappeared.

I turned back to the cage, puzzled. Wasn't that part of the fun of it all, meeting challenges? The gorillas certainly did not avoid difficulties. In fact, Lulu sometimes forced Patty to struggle for something. She gave the baby challenges. Patty even competed for food with her parents. It didn't seem to hurt her but, rather, made her a bit more aggressive. She never went hungry but simply headed for another piece of orange or another slice of apple.

She struggled to hang onto her mother while Lulu and Kongo played, made love, or even fought. In the narrow existence of the cages she struggled to explore. She struggled to be free of Lulu, to investigate a bit of sunlight that splashed onto the floor, then

132

changed her mind and struggled to be with her. She climbed up over Lulu even as her mother turned away or unclasped the fingers of one of her hands. She screamed when she wanted to nurse and tried to bite when she was restrained—obstinate, insistent, and totally active.

As spring approached, her physical abilities had already blossomed. Now her temper and her will were developing quickly, and temper tantrums became a part of her behavior. She wanted what she wanted when she wanted it. And she struggled and fought to do what she wanted to do. Frustration entered her life. It was not a negative force, but a positive one. The frustrations that Lulu imposed on her infant seemed to increase Patty's desire for independence.

Lulu often seemed to be holding Patty back. She held onto her firmly; Patty scrambled vainly to escape, fiercely clutching at air while her legs churned against the ground. Lulu sat calmly, almost absentmindedly chewing on the chain at the front of the cage, while Patty flailed arms and legs, struggling with all her might. Then, suddenly, Lulu would let her go. As soon as Patty got her way, the temper tantrum vanished and the writhing stopped. Within less than a moment she was crawling around unconcernedly, perhaps heading for a bit of orange less than a foot away.

I admit that the sight of the infant's being restrained looked cruel. And it did not surprise me that many people watched with horror. But what so many people regarded as mistreatment of Patty Cake might well have been a natural part of her training. It certainly did not harm her, nor did it keep her from doing what she wanted to do. The frustration that it provided could have been, in a small way, preparation for the future.

There is no strife-free existence. An animal must be allowed to become strong, even competitive. This playlike struggle with her mother was an introduction to challenge and, I thought, would help establish her acceptance of whatever would later come her way. In the cages in front of me were the order, discipline, and strength that would make Patty Cake a beautiful gorilla.

One day H. Bradford House, curator of mammals of the New York Zoological Society (the Bronx Zoo), stopped by just as Lulu was tossing Patty into the air in a series of whirling somersaults.

He watched with growing concern as Patty freed herself from her mother's determined exercise and headed for the bars. Nonchalantly Lulu reached out and grabbed a leg. The usual performance began. There was the writhing and the sounds of angry annoyance and the twisting and the turning. Brad shook his head at the sight of the struggle.

"She [Lulu] makes me so mad," he said. "She's an impossible wench."

A few minutes later Patty was happily sliding down the bars like a fireman for the first time. But he was gone.

CHAPTER 20
The Six-Month Birthday Party March 3, 1973

It was March 3, and Patty was six months old! I waltzed happily into the Lion House to find it a very busy place that morning.

"What's going on?" I asked Luis as he hurried past me in the keeper's walk, his mop trailing behind him on a newly washed floor.

"The press is coming," he said. "The big shots, they want to film Patty Cake. She's going to be on television."

I looked at the newly polished plexiglass. It had seldom been so clean. By the time the media began to arrive the entire house was spotless.

The chain clinked against the doors again and again as people arrived. They came carrying leather boxes and cases filled with television equipment. People with note pads wandered around, asking the keepers questions. Lights were tested, flashing on and off, and the sound of cameramen testing the microphones for sound quality filled the house. Fitz had slipped in quietly and stood leaning against the railing, barely noticed by anyone else. Charlie, the tiger, began his long, low moan as he waited for his bath. A few people gathered in front of his cage to watch him revel in the stream of water that Luis hosed up and down his spine.

Kongo came to the front of the cage, alert to each movement of

135

the growing crowd. Lulu moved to the rear, holding Patty tightly against her stomach. She stood shifting her feet, unsure, near the doorway to the number two cage. Her gaze went nervously back and forth from Kongo to the scene in front of her cage. I doubted that Lulu would let Patty go or that she would even come close enough to the front of the cage to allow clear shots of the baby.

The lights were turned on. The animals blinked rapidly in the astonishing brightness. The glare bounced off the plexiglass, blinding us.

"We can't film through that!" The cameraman pointed in disgust at the plexiglass. "We're going to get nothing but glare. Can we go behind it?" As he spoke, he strolled into the keeper's room and then into the walkway. Kongo charged, and the man jumped back, splashed with urine.

"What the!"

"Oh, boy, I'm not going in there!" A reporter turned to Fitz. "Look, we can't film in here. It just isn't going to work. Can't we get the animals outside?"

Since the day was sunny and warm, Fitz gave permission for the apes to be allowed outside for one hour. It would be Patty's first day out since the fall. There was a new bustle as the men packed up their equipment and began to transfer it outside.

"Hi."

I looked up to see Paul Dandridge of CBS. I recognized him immediately from having watched him on the evening news.

"It will take sometime to set up," he said. "Would you like a cup of coffee or something?"

"I'd love it." And we strolled over to the zoo cafeteria. I settled at a table while Paul got some tea.

"How did you get here?" he asked.

"It's a long story," I said with a laugh.

"I'm really interested," he said. And he was. By the time I was halfway through my tea I realized that he was writing down what I was saying.

"Tell me about the gorillas," he said. And I was off a mile a minute. I began to explain the emotional relationships that the apes seemed to have developed among themselves and the ways in which Lulu was bringing up her baby. He leaned forward with interest. His face was thoughtful.

136

"You know, I'd like to interview you," he said.

I stared at him. A nervous jab caught me in the stomach. "All right"—I laughed—"it'll be fun."

When we returned to the Lion House, the crews were almost finished setting up their gear. Paul headed toward his crew to put his notes together and write a monologue, perhaps decide which questions to ask me. It was an exciting thought. All the reporters seemed to be mumbling, staring into space as they rehearsed.

Police barriers had been set up while I was gone. They formed a large semicircle in the area in front of the gorillas' cages. Behind them perhaps 100 people waited for the sight of Lulu and her baby. Men laden down with heavy cameras climbed gingerly over the railing. The reporters were still getting last-minute information. Fitz watched them silently, leaning back against a tree next to the cages. He stepped forward every once in a while to tell someone not to get too close to Kongo, who was about to rush into his cage, or to gesture that another reporter's position was all right. Finally, all was ready, and the door to the inside cage was opened for the first time since November. The crowd grew quiet in anticipation.

"Here she comes . . . ready."

"Shhh, here she comes."

"God damn it, I'm at the wrong door." There was a rustle of hurried movement as someone shifted his camera to another position.

Lulu's eyes, tentative and searching, peered out at us from the dark cavern of the cage. Slowly and delicately she came out. Patty was hidden, clinging to Lulu's stomach (a position now reserved for times of suspicious circumstances). The whirring sound of the cameras sent Lulu hurrying back to the doorway. As the cameramen shut off the cameras and the sound ceased, she turned toward us again. Then went in.

But Kongo was not there, and she stood in the doorway, excited and undecided. Richie and Eddie called softly to her, and hesitantly she came out, holding Patty fast beneath her. The cameras began to whir again as Lulu crossed the cage to the keepers. The men petted her and spoke to her, and within a few minutes she felt secure enough to shift the baby into an exposed position on her back. Patty clung and peered out at us from her mother's hair, her eyes wide and round. She stared at the people and the equipment, then

137

clutched her mother more closely, seeming to sink into her, merge with her.

Lulu was no longer apprehensive. This place was not strange to her. Every spring, summer, and fall the outside cages had been her home. She knew the light and the space and the air of the out-of-doors. She put Patty down.

But Patty had not really known this place, although she was born in an outside cage; this new space, nine feet square, was virtually unknown to her. There was no smell or lions or tigers there, the animal smells were diffused in the air. The air itself, light and breezy, the sunshine, the depth of the space, and the sight of bushes—all these things were new to her. Then there were the people doing strange things very close to her cage. There was the equipment, most of all the cameras, whose long, projecting one-eyed lenses were aimed at her and followed her. And little fluttering, twittering birds hopped on the floor of the cage around her. When Lulu put her down, Patty clutched her mother's ankle and wrapped her arms around her leg. She crawled with her mother, pulled along as Lulu moved. Her rear dragged low toward the ground. It was a sign of fear in the gorilla, and I had never seen it before in the six months of Patty's life.

She stretched her arms up, tugging on Lulu's leg. She tried to climb onto the familiar comfort and safety of her mother. Lulu looked down at the upturned, pleading face and lifted her daughter onto her stomach for a few minutes. When she again put Patty down, her daughter cringed against her leg and clung again. Lulu walked across the cage, and Patty, clutching her ankle, lurched forward with each step. When Lulu sat down, Patty scrambled into her lap and pressed herself against Lulu's body. When Lulu put her on the floor, she screamed. Lulu swooped her up to her breast, and Patty sucked ardently. Lulu picked up a leaf and began to chew on it.

A huge crowd had gathered behind the barriers. Children, anxious to see, timidly crept forward for a better view. Men and women strained to see past each other, oohing and aahing each time Lulu nursed Patty or picked her up or flung her onto her back. And when Lulu carried Patty over to the wall between the cages and peeped through the slits at Kongo, they marveled and whispered to each other that "she knows where he is."

138

One voice stood out. "Look, sweetheart, see the little baby? Isn't that cute?"

I turned to see a woman holding a little white poodle in her arms. I turned back to the gorillas. Suddenly I felt a woosh of a movement under my feet and the poodle streaked through the bars of the railing and darted toward Lulu's cage. She started toward him, staring at the eager happy puppy and had nearly reached him when he hopped to the ground and headed toward Kongo's cage, stood up against the cage and wagged his tail as Kongo came toward him. I jumped, tripping over a microphone wire.

"Fitz! The DOG!"

Just as Kongo's hand reached out toward the dog Fitz made a flying leap. The dog squirmed in his arms.

"Where does this belong?" he said, trying to contain the wiggling animal. Before I had a chance to answer, the owner emerged from the crowd and took the wiggling mass of snow-white curls from him.

"Lady, please, keep your dog on a leash. He coulda got killed just now." The woman stared at Fitz blankly for a moment then melted back into the crowd.

Fitz looked at me and shook his head. Then we both turned back to the cage.

Eddie was tossing some food to the gorillas. Patty became a little more independent. As Lulu regurgitated the Italian ice someone had given her, Patty followed her mother and also bent over it. Then she timidly left her mother's side to climb. She climbed onto and then over her mother's foot, which was propped up against a bar, looked back at Lulu, and touched her. Slowly she began to climb away from Lulu. But as she did, one of the cameramen leaned too far forward, and suddenly Lulu went after him. As her left hand grabbed at him and his camera, her right hand went for her baby, pulling her from the bars and pushing her behind her. Patty squatted there, clinging to Lulu's back, her eyes once more wide and apprehensive. She sat absolutely still. When Richie came over to talk to Lulu, she picked up her baby and turned around, hiding her in the hollow circle formed by her body. As Richie scratched her back, her head repeatedly swiveled around to look at the queer foreign instrument which had come too close to her baby.

Little by little the gorillas relaxed as the cameramen moved back from the cages. Soon Patty was following Lulu around the cage, venturing out to climb and explore, looking back at Lulu for approval and reassurance as she moved away.

Paul Dandridge headed toward me. "If it's all right with you," he said, "I'd like to get that interview done now." It was very exciting to be interviewed, and a little strange. I felt self-conscious as the cameras focused on us and Paul began asking questions to which he already knew the answers. He looked and sounded as if they were absolutely spontaneous.

"Look, can you move over a bit? The sun is moving." The cameraman interrupted us in the middle of a sentence.

"Sure, how's this?" Paul moved us over a bit. I realized that this was just a rehearsal, and I could relax. It was fun. My voice ceased to shake as I answered the rest of his questions. He grinned at me.

"That was fine," he said. "It's going to be great. Ready to do it again?"

His voice sounded just as spontaneous the second time as it had the first. Suddenly it was over.

"Thanks very much," he said. "I'll be looking for your book. It ought to be interesting. Good luck."

I turned back to the gorillas only to find someone else asking if he could interview me. I felt like a pro.

It was time for the animals to go in. As the door opened, Patty scurried toward her mother. Simultaneously Lulu strode toward Patty, and they met halfway. Patty found herself once more deposited on Lulu's back and carried swiftly into the familiar cages. There Lulu put her down. Patty curled up on the floor and fell asleep. She always seemed to fall asleep after any excitement or new experience. Lulu did not disturb her but lay down next to her. She watched her for a moment, then leaned over and licked Patty's head. When Patty did not stir, Lulu, too, settled down and went to sleep.

CHAPTER 21
I Want to Play with Daddy
Sixth Month

Patty had discovered the diamond-shaped grating that covered the bars at the edges of the cage. Intrigued, she would hang onto the bars next to it and lean over to examine it closely. She sniffed it, licked it, ran her fingers over it, then poked the tip of one finger into a little diamond-shaped hole. It just fitted.

One day at the beginning of March she saw something move from the corner of her eye. She turned to find Kongo watching her. She could just see the top of his head and his wistful eyes peering over the top of the metal plate wall. She broke off her play and stared at him. At that moment she spotted Lulu coming toward her in a matter-of-fact way. Patty climbed down the bars and scampered away. But contact had been made.

Each time she headed for that corner Kongo seemed to be there. And each time Patty stopped to stare at him. She began to look for finger grips in the grating, but the holes were too small for her to clasp. She was eager to go to him, insistent on finding a way, searching for some way of crossing that vexing space of the corner. Meanwhile, Kongo waited. Soulfully he gazed through the bars, his eyes fastened on her. Now, as she stared at him, he wiggled a finger at her.

For a moment she hung suspended. Then she stretched her right arm across the triangle of the corner. She stretched farther than she ever had before. She caught hold of a bar on the other side of the corner. In another moment she had swung across the triangle of space.

She lost her grip and dropped, hanging on to the bars with one hand while the other swung free. She stretched up to grasp the bar again. Then, pulling on the bars and pressing her feet against the wall, she raised her body until she could peer over the wall. Her eyes met Kongo's on the other side.

Suddenly Lulu's hand engulfed her and pulled her away. She was deposited unceremoniously on the floor. Kongo sat unmoving, watching, on the other side.

Patty wasted no time. Having discovered the route, she rushed up the bars and swung more easily across the corner of the cage. She stared into her father's face. Kongo touched her; one great sausage finger stroked the tiny hand that held onto the bar. Gently he poked the stomach that pressed in between the bars. When Lulu took her away, neither of them protested. But Kongo stayed at the bars, watching her, and Patty peeked up at him from her mother's arms.

From then on Patty was fascinated by him. Whether it was a result of her mother's insistence that she not go to him, the newness of swinging freely across that triangle of space, or Kongo's silent, ever-watching presence and obvious invitations, Patty would not stay away from him. Every time Lulu's back was turned Patty scurried to the corner, gleefully ran up the bars, and swung across. Kongo was waiting; he wiggled his fingers at her, enticing her, his eyes watching her expectantly. She would grab a tremendous finger and pull on it or reach between the bars to touch his face. And he would nuzzle and poke her as she toyed with his ear. They played together eye to eye until Lulu came to take her away.

CHAPTER 22
The Accident
March 20, 1973

From my diary for the morning of March 20, 1973:

She is so sweet, reaching for these familiar faces with no fear, coming close to the men as they come to pick her up and lift her playfully on the other side of the bars that separate them. And now to, Fitz, whom she does not know as well, her black palms and fingers spreading as she reaches out to touch his face. And of course, he lets her. All trust, this little baby, with no reason to be afraid.

"Susie! Please, do not be frightened."

I looked up quickly into Andrée's fear-filled eyes and dropped my sandwich.

"What's wrong?" I was already on my feet.

"Oh, Susie, there is an accident."

I scooped up my purse and headed out of the cafeteria. "What happened?"

"Ah . . . Patty was going to Kongo, you know, the way she does . . . to play with him— "

"What happened?"

"When Lulu took her away—her arm—she struck the metal bar.

There is something wrong with an arm." Andrée caught her breath in a little sob.

I pulled open the door to the Lion House and rushed in against a stream of people. I ran down the corridor and pushed my way through the crowd in front of the cage.

I could see at a glance that one of Patty's arms was limp and crooked. There was something terribly wrong. And Lulu knew it. Patty's hand did not come up to grasp her hair as usual. There was some movement of Patty's shoulder as she tried to lift the hanging arm to her mother, but she couldn't do it. Lulu took hold of the limp hand and tried to help her daughter hold onto her hair with it. But when she let go, the arm fell and dangled. Then Lulu lifted the baby up by that injured right arm, putting her hand over the place where it was broken. Pacing, confused, the mother tried to make the hand hold onto her, placing it against herself and then letting go. But the arm dangled uselessly; Patty clung tightly to her mother with her left hand alone. Her bright eyes looked up into Lulu's face as her mother carried her. There was not a sound from the infant— not a cry, not a whimper. She just clung silently, her head tilted back.

Lulu paced nervously. She began to race around the cage, climbing the bars, becoming frantic with fear. As always when she was afraid, she began defecating with diarrhea. She looked out at us and then put the baby down and picked her up again. She turned Patty over and over in her hands, examining her daughter's body for some sign of an injury, but she found nothing. Perhaps if she had seen an open wound, it would have calmed her. She would have examined it and licked it, caring for her baby. But this injury was hidden from her view. She seemed confused, not knowing what was wrong and what to do. For the first time since I had known the gorillas, I wished that she could understand words; I wished that I could tell her what was wrong.

In the meantime, the shocked and worried keepers began to clear the house of visitors.

Kongo sat motionlessly watching Lulu and the infant from the neighboring cage. Now the men decided to put him outside so that they could open all the inside cages to Lulu and her baby. They

144

knew that they would have to separate Patty from her mother for the first time in her life.

For once Kongo gave the men no trouble, and he bounded out into the fresh spring air. The door closed behind him.

All attention now centered on the mother and the baby.

I felt sick as I watched the gorillas, knowing that Patty was now in pain.

Patty continued to try to lift her broken arm, raising it slightly, but she was unable to control it. Lulu took the baby up to the window with her and strained to look outside, trying to see Kongo, I thought. She came down again, paced, climbed up to the crossbar, then down, restless and nervous.

Fitz ran in. He rushed to the cage. But there was nothing that he could do until the doctor came. He leaned back against the far railing and watched the gorillas silently, his eyes hidden, as usual, behind dark glasses, his face pale, tense, and drawn.

Dick Berg approached. "Someone could feed Lulu and try to make her put the baby down," he said. "Then I could go in and take Patty out."

Fitz looked at him incredulously. "You've got to be kidding," he said. "She'd kill you!" He turned back to the cage.

Dick, morose, drifted toward me. "I wish he'd let me go in," he said mournfully. "It doesn't matter what happens to me. Patty's the important one." Miserably he went to get an Italian ice for Lulu, wanting to do something, however useless or irrelevant.

Lulu paid no attention to any of the bribes. They tried ices and grapes and practically everything else she liked. They talked to her and tempted her to put the baby down and play. But there was no way that she could be enticed to leave her infant. There was nothing that we could do, so we just stood there silently, each of us alone.

Ed Garner finally arrived. He took one look at the infant clutching her mother's leg and confirmed what we all already knew. "Well, we'll have to tranquilize her. We'll have to take the baby away." He sent one of the men for his equipment—the syringe, dartgun, darts, drugs, and cloth.

"I really don't want to have to shoot her," Ed said. "They keep

145

moving, and it's hard to get a good shot. We want to put the baby in as little danger as possible.'' Fitz stared into the cage and nodded. "Let's give her small doses by hypodermic, if we can, not to knock her out completely, but just to relax her enough to put down the baby.''

Dick Berg came over to reassure me.

"It's only a sprain,'' he said. "All the keepers think that's all it is.'' He looked at me hopefully, and I nodded in agreement. But as I looked at Patty, I could see the break in the misshapen arm and, below it, the rest of the limb swinging freely.

Lulu set her down on the floor, and Patty lifted her chest off the ground and tried to support herself on her arms. She tried to crawl on that broken, crooked arm. The arm below the break twisted so that her wrist turned inward on the cement. Whimpering, she tried to creep to her mother. It was the only time that she had cried so far. She leaned weakly on the arm before it collapsed under her, and she dropped slowly to the floor, crying. Looking at Lulu a few feet away, she seemed to be pleading to be picked up.

And Lulu came for her.

I could no longer bear to watch helplessly as Lulu floundered in confusion, examining and comforting her child. I turned away from the cage and headed down the corridor. Questions ran through my mind: would Patty be afraid of Kongo? When would they put her back with them? Would it ever be the same again?

I found myself standing in front of the baboons. Big Boy was masturbating, watching himself with great interest, pulling on his long pink penis. When the semen came, he squeezed it out and licked it off his fingers. I realized that I was witnessing another rarely seen incident, that there was still another world outside the cage in which Patty and Lulu were living through their agony.

On my way back to the gorillas, I passed the two old gorilla ladies. Caroline was standing against the wall, straining to look through a peephole into Lulu's cage. One of her arms was shaking nervously. She chased back and forth from one peephole to another as Lulu moved in and out of her view. Joanne, on the shelf above, looked in on them, grunting, They heard the sounds of Patty's whimpers coming from the next cage.

We tried to keep out of Lulu's sight, thinking that less activity

146

around her cage would calm her. An hour passed as we stood waiting in silence. Where was the doctor's equipment? Why was it taking so long to arrive?

Richie told me that Patty was using the injured arm, that gradually she was beginning to cling to Lulu with both hands, but I saw that it really wasn't so. The keepers cared so much for these animals that they were trying not see reality. Only Luis, sad and quiet, went about his work of cleaning the other cages, knowing there was nothing he could do for Patty.

I heard a familiar sound behind me and looked back to see the cats Princess and Janet, licking and pawing at the door between their cages. The lions and tigers were restless; it was almost time for them to eat.

People were knocking at the door and calling to be let in. The press was arriving en masse. Norman (Skip) Garrett, of the Public Relations Office of the Parks Department, slipped into the house. He walked toward Fitz and Ed apprehensively. "We've got a problem," he said. "You know the press is outside. They're clamoring for information. They want to know what's going on in here."

The two men looked at each other, resigned to the inevitable publicity. Ed sighed. "Just tell them that we're going to try to take the baby away from Lulu in order to do a thorough examination."

As time passed, Skip ran back and forth with bulletins. "They want to know if they can film it when you take Patty away from Lulu," he told them. Fitz looked at him, startled.

"No one is to come in here!" he ordered quietly. "No one!"

At last the equipment arrived, and there was purposeful activity in the house. Ed began to prepare the hypodermic; the men formed a little circle around him. Fitz and Ed had decided that Richie would try to give the shot to Lulu. There were two reasons. First she knew, trusted, and liked him, and secondly, he was used to giving transfusions to his young son, Chris, who was, tragically, a severe hemophiliac. Richie listened carefully to the instructions. He positioned himself by the bars and called for Lulu to come to him. Instead, she left Patty on the floor at the front of the cage and moved back, staying out of his reach. Patty lifted her chest off the ground and again tottered forward, whimpering for her mother. Her arm was dragging, long and limp, across the cement floor. Tentatively Lulu came toward her. As Richie stood by the bars with the syringe in his hand, she scooped up the baby and ran to the rear of the cage. She stayed there, eyeing him suspiciously.

We waited and waited, but each time Lulu came forward, she backed off again skittishly. Richie stood with the syringe hidden behind his back. Finally, she came to him; he waited until her head was turned away, then tried to position the needle. Two or three times she shifted away at the crucial moment. She would not come near him again.

They decided to use the dart gun.

Robert Beach, one of the head keepers, stood holding the gun in front of the home cage, waiting for his chance. It was imperative that he hit Lulu, but not the baby. He would have to be quick and sure, for if Lulu turned at the decisive moment, how could he judge where the baby would be? He also had to make sure, somehow, that Lulu would not drop the infant. This was deadly serious; timing was of absolute importance.

He leaned the dart gun against his left arm in a marksman's stance. Lulu moved suddenly, and he lowered the gun. Again and again he raised the gun; as Lulu turned or swung away, holding the baby close to her, he lowered it.

At nine minutes to two in the afternoon, as Lulu paused momentarily on the crossbar, he shot her.

It was a beautiful shot in the left thigh.

She screamed! Her mouth opened in a horrible grimace, and she swung to the floor. The screaming went on. Pressing Patty against her belly, she bolted into the next cage and up onto the crossbar for safety. The dart hung grotesquely from her thigh. She crouched, panting, her terror-stricken eyes glaring out over the board that hid and protected her from us. Through an open slit in the wall, I saw her examine the dart. She yanked it out of her body and hurled it to the floor. Supporting Patty on her back, she swung wildly around the cage, put the infant down, and left her to run from one side of the cage to the other. Alone, Patty seemed oblivious to the terror. She managed to crawl to a piece of an orange lying nearby and put her face down over the fruit to lick it. Then she began to whimper. When Lulu heard the little cries, she came to her.

The drug should have taken effect within twenty minutes, but at

148

the end of that time Lulu still swung from place to place. Patty tried to follow her from below, making pitiful little birdlike cries as she stumbled around the floor.

A half hour passed, there was no sign of the drug's taking effect. Richie prepared to make another attempt to inject her.

He stood by the bars, talking to her, showing her bits of candy. She skirted him, her eyes watchful and suspicious. He hid the syringe and showed her his empty hands.

"Look," he said, "I don't have anything." Lulu looked into his eyes as he spoke. "Come on Lu, let me scratch your back," he tempted soothingly, "that's a girl." Haltingly she came forward and sat down by him. She turned away slightly, and he scratched her back with one hand. Slowly he reached back for the syringe with the other. Suddenly he injected her. She sprang away from him and began to run, looking back at him and the dripping needle that he held. He had managed to inject most of the one-quarter of a cc of medicine. It was 2:18 P.M.

On the other side of the cage now, Lulu glared at her trusted friend Richie, her eyes filled with suspicion and fear. When she picked up Patty, the infant began to nurse. Lulu lay down as far from us as possible. But she did not stay still. As the drug began to work, she began to swing again. She went from bar to bar, but we could see that she was a little groggy. The men tried to tempt her down off the crossbar, where she finally perched with her baby. She came down and began to pace, and she seemed to lose her balance a bit; she took Patty up to the safety of the crossbar again. She swayed a bit; suddenly the baby fell.

"Oh, my God," whispered Ed. And Fitz, pale and shaking, put his hands over his face.

Patty landed on her back with a thud and a sudden sharp, high-pitched scream. She had not fallen onto her injured arm. Lulu slipped uncertainly down to her, staggered a little, picked her up, and sat down jerkily. She turned her over and over, then held her to her breast. Patty nuzzled her, trying to nurse. Her eyes were open, and we saw her good arm reach up to touch her mother's face.

The men were trying to move Lulu out of the home cage and away from that crossbar where she felt so safe but was really in

149

danger. First, they would have to trap her in the number two cage. But Lulu was frantic and confused, and she would not leave her favorite cage. Twice she was tempted into the second cage. But as Richie began to lower the door, she put her hands under it and tried to lift the heavy metal until he opened it for her again. Twice she lifted the door, then ran back into the safety of the home cage, holding Patty against her and panting. But finally, the third time, she was not quick enough, and suddenly the door clanged shut. Now there was no place for her to go but into the number three cage. Within moments she was trapped there.

Waiting, waiting. Interminable waiting. We heard the panthers growling and the baby baboons screaming at one another. We heard Charlie's low moan and the lion's familiar rhythmic roar. And we heard Kongo outside, pounding on the door. We could see just the top of his head as he stood on the sill outside and tried to look in through the window. And we heard him crying.

The day went on and on, and Lulu still seemed to be alert.

We could not reach the baby.

From my diary: 4 P.M.:

Robert Beach is going to shoot her again. She sees him and that black, heavy gun, and she is beginning to scream, swinging, backing away from him, moaning with fear. But there is no place she can hide. She is completely exposed, trapped.

Caroline and Joanne, looking through their peepholes and hearing Lulu's shrieks and sobs, suddenly begin to scream. Caroline is warning us with threatening grunts from deep in her throat.

Lulu backs away from Beach again. She screams, moans, long, low, tortured cries of fear, and I, too, am afraid for her terror, afraid of the sight of that horrible gun. I, too, shrink from them and creep back to where I cannot quite see. She has never been hurt or betrayed by these people before today. I cannot watch or listen to her; it hurts me to hear her terror.

She screams.

There is the shot, and the screaming! The screaming . . .

Luis was already taking the bolt off the cage door, preparing to open it for Fitz, who would go in and take the baby. Lulu was frantic. She left Patty, turned, picked her up, and, confused, put her

150

down again. Patty shrieked, too, when Lulu left her. And I realized that she had gone from a whimper to a cry to a scream today in a progression of unknown fear for the first time in her life. She tried hard to follow Lulu, but now her mother paid no attention to her. The drug was finally working. Lulu held onto the pole, trying to stand upright. Suddenly she fell, brushing Patty as she staggered. Patty tumbled over onto her left arm, then struggled to right herself. She looked up at her mother with fear as Lulu again lurched over her, looming up, unsteady above her. And Lulu, her eyes still on her baby floundering feebly on the floor, collapsed next to the pole. She pulled herself up, leaning weakly against it. She turned toward the wall and stretched toward it, staggered against it, and reeled toward the door, trying, I thought, to reach Kongo. She fell and pulled herself up, dragging herself over to his cage, looking back at Patty in the middle of the floor, abandoning her. As she made her last effort, she lurched through the open door, pitched forward, and sprawled there. Patty was screaming. *Screaming*!

Luis swung the cage door open, and Fitz leaped in. At 4:10 P.M. he brought Patty out.

He hurried down the keeper's walk, cradling her in his arms; she looked far larger than she had ever looked in her cages.

"Where's the blanket?" Fitz asked. Someone rushed forward with it, and he gently wrapped it around her, speaking soothingly to her all the while. She lay in his arms, swaddled in the green and black plaid blanket, looking up at him. Her eyes were large and bright; she gazed trustingly into his face. She raised her left arm and stroked his cheek with one little finger.

He rushed to the rear door of the house. Ed Garner followed quickly. Soon the door closed behind them, and they were gone. Patty was gone. The keepers dispersed silently, and the house was quiet.

I went back to Lulu.

She had been hanging onto the bars between the cages, trying desperately to hold herself up when she saw Fitz enter her cage. And when he picked up her infant, she had clawed at the bars, flopping over, her limbs uncontrollable, falling, yet frantically reeling along the wall toward her baby. She saw him take her infant from the cage. And when she reached the door between the cages, she

151

had fallen onto her back for the last time. Her eyes were wide with fear. They followed Patty as long as they could. Now she lay still. Only her head rolled silently back and forth on the floor.

"Lulu," I cried to her. "Lulu, it's all right. She's safe."

I heard the clink of the chain as the door opened. Paul Dandridge of CBS poked his head through.

His face was strained and white.

"Susan. You've been here all this time?"

I nodded.

"How is she? What happened?"

"I'm sorry, Paul. I can't tell you anything." Suddenly I felt tired.

"Susan, you know me. You can tell me." He looked at me anxiously.

"She's got a broken arm," I said. "That's all. She'll be all right. Really, that's all I can tell you. She's going to be fine."

He nodded and disappeared behind the closing door.

I turned back to the cage. Everyone else was gone. Now it remained for the associate director of the zoo, John Kinsig, and Luis to bring Lulu into the cage from which her baby had just been taken. Luis spoke to her gently as they lifted her and brought her in. She flopped over helplessly as they laid her down, and her arm fell to the floor as if it were boneless. Luis bent over her and patted her head once more to comfort her; then the men left the cage. She was alone.

They closed the door between the cages, then let Kongo into the house.

I had almost forgotten Kongo. A zoo regular told me later about him. Kongo had been outside and had heard Lulu's screams. He could not reach her. At the beginning he had paced up and down and seemed to be looking for his mate. When she did not join him in the outside cage, he climbed to the crossbar to look through the window for her. When he heard her first screams, he went wild. He charged around the cage, ran to the doors, and tried desperately to open them. He vaulted up to the windows again and watched her from there. He must have seen her agony and confusion.

For all those hours he tried to get to her, putting his hands under the doors, straining to lift them. We inside heard him banging and

152

crying as he became more and more frantic and afraid. He was at the windows when the second set of screams came from inside. He must have seen her shot again. And in frustrated fear and helplessness, he defecated in diarrheal spurts. His moans grew louder as he tore at the unyielding windows and door.

When the door finally opened, he rushed in to Lulu. But he was stopped by the bars. He looked down at her, sprawled out on the floor, her head rolling feebly back and forth against the cement. He saw her as she tried to crawl to the pole and crumpled onto the floor. Her eyes were round, filled with drugs, horror, and helplessness. And there were bloody cuts on her.

Kongo stood there, watching her silently.

I did not want to leave them that night. I wanted to stay and speak soothing words to comfort Lulu. But Luis touched me softly on the back. He was sad and tired, too. "Whatsa matter, Susie? She'll go to sleep soon. She'll be all right."

I could not stay. One by one the lights went out, and I gathered up my materials and slowly walked down the familiar corridor. As Luis and I passed Kongo's cage, I could see him still standing there, quiet and motionless in the darkened room, a mountainous hulk as he kept his vigil, as near to Lulu as he could be.

And then we passed Lulu, alone now.

CHAPTER 23

To the Hospital
March 20, 1973

It never occurred to me to follow Patty Cake. I assumed that she was on her way to care and sleep and that after all these waiting hours she would soon be snug and safe.

There was no hint of the commotion that was, at that moment, taking place behind the silent buildings. For Patty was not on her way to the hospital. There would be hours of further pain and confusion before she would be cared for.

The newspeople had waited outside the Lion House for Patty to appear. They milled around the eastern door, talking, joking, and relaxing. Fitz and Ed slipped quietly out the back, unnoticed.

Fitz carried Patty in his arms, walking urgently toward his office. Ed followed, almost running to keep up with them. His eyes were glued to the little gorilla. At the Lion House someone happened to glance that way.

"There they are!"

It was as if someone had thrown a stone into the midst of a school of goldfish. The two dozen people scattered, scrambling for their equipment. They sprinted toward the two anxious, hurrying figures.

Breathless cameramen encircled Fitz and Ed, moving with them, sometimes running backward to get better shots. Some men trailed out behind, their heavy cameras weighing them down. Others raced ahead to the office to meet the entourage as they approached the door. Someone put a microphone in front of Ed and began to question him. People shifted about, nearly dancing around them.

There was a confusing barrage of questions, demands, orders, and moving, shifting, excited people. At the eye of this chaos were two worried men and a small, confused gorilla.

As they approached the office, someone opened the door. Then it closed behind them. Fitz carried Patty down the hall to the rear of the building and into the zoo kitchen. Gently he laid her on the table. As Patty gazed up at them, Ed carefully unwrapped the blanket and bent over her. He touched her injured arm and could feel the two separate pieces of bone grating against each other. She made no movement or cry. She just looked up at him steadily with big, round eyes.

"It feels like a clean break," Ed muttered. "Let me call the hospital. I'd better make some arrangements with them to use the X-ray department. We don't have any X-ray facilities in the animal care unit."

He disappeared into Fitz's office to make the necessary calls. Fitz stood by Patty, stroking her softly. She seemed unafraid, and she clutched one of his fingers with her left hand and stared at him.

Ed spoke on the telephone to Dr. Felipe Montoya, the pediatrician at New York Medical College who would make the arrangements for them.

"Call me back in five minutes," Dr. Montoya said as soon as he heard what the situation was. But that was impossible to do, for as soon as the news of the accident was out, the telephone rang steadily. The office was in a state of tumult. Photographers were knocking on the door; reporters poked their heads in to ask questions.

The Parks Department public relations people began to arrive.

"Let's get the commissioner to take Patty to the hospital in his car! Boy, will that look good!"

Someone ran to the phone and called the Commissioner Richard Clurman's office. They no sooner put the telephone down than his office called them back. They were told no. The commissioner

156

would not take Patty to the hospital. Mr. Clurman was not going to make a publicity stunt out of the animal's injury.

The phone rang again immediately. It was a news service wanting to know what was happening.

Skip Garrett ran back and forth, trying to get information to pass on to the restless press outside. Each time the door opened a babble of voices was heard. Each time it was louder.

Ed ran back and forth, trying to make his calls, stopped by people asking questions. He rushed back to Patty, who lay quietly in the midst of the hubbub, still clutching Fitz's hand.

Finally, Ed gave up. "Come on," he said to Fitz. "We'd better give them something to go on. Maybe they'll leave us alone if we do."

Fitz agreed. "Tell them we'll have something to say shortly," he told Skip.

The telephone rang again. Finally, the news came that all the arrangements had been made at the hospital, and they could leave anytime.

"Someone get my car!" Ed called to one of the keepers. "Bring it around to the kitchen door." He bent over Patty to wrap her in the plaid blanket.

"We'll talk to the press on our way out," Fitz said. He turned to Ed. "That way we can make it as short as possible and get out of here."

The two men with the quiet baby gorilla stepped outside. Immediately they were surrounded by the newspeople. They stood next to the car and gave a brief statement. In a minute it was over.

Ed jumped behind the wheel, and Fitz climbed gingerly into the passenger's seat with Patty cradled in his arms. As they drove off, the NBC crew raced for their car.

They sped down the winding roads through Central Park, past the zoo, the baseball fields, the Metropolitan Museum of Art, and the reservoir. All the while they could see the news car following them on the empty road. As they pulled up to the emergency entrance of the hospital, the NBC car came alongside.

One of the crew rolled down his window. "Will you wait five minutes? We want to get set up!"

157

Ed, nonplussed, glanced at Fitz.

"Are you kidding? You've got to be kidding!" Fitz responded. He turned to Ed. "Let's get parked and get her in there."

But by the time they parked the car the press had already grabbed their equipment and were ready to shoot. They came running, following the two men and the little injured ape into the hospital. The cameras clicked and whirred. When the elevator door opened, they could see a patient inside, lying motionless on a stainless steel table. He was covered with a white sheet. The attendant by his side took one look at them and cried out, "Oh, no, you can't bring that animal in here!"

Even as he was finishing the sentence, Ed was already waving Fitz down a corridor. "This way!"

Marjorie Margolis of NBC called after them, "Where are you going?"

"X ray!"

While Ed and Fitz hurried through the halls to the other side of the building, waited for the elevator, took it upstairs, and then retraced their steps on a higher floor, the television crew merely stepped into the first free elevator and zipped upstairs. By the time Ed and Fitz arrived in the X-ray department the crew was already there, waiting.

Other newspeople began to arrive. Since Ed and Fitz were unable to get away while the X-ray room was being prepared for them, they gave another interview. At last a technician leaned out into the hall from the X-ray room. "We're ready for you," he said. "You can bring her in now."

Fitz took Patty in, laid her on the table, and gently unwrapped her again. The crowd began to swarm in behind him.

"I'm sorry, you'll have to leave the room," the technician informed them. "No one but staff is allowed in here. Radiation."

At the word "radiation" they left hurriedly and clustered in front of a window that looked in on the action. Even Fitz was asked to wait outside. Standing apart from the others, he leaned tiredly against the wall, still holding the green and black plaid blanket. His eyes were closed.

Inside, the doctors, gowned with the heavy lead-lined aprons which would protect them from the X rays, bent over Patty. Ed held her down and positioned her. The huge machine swung over her and lowered to take the pictures. Ed shifted her into another position, and again the machine was placed precisely over her arm. And again and again. There was not even a whimper from her. She simply kept her eyes glued to Ed's face.

When they were finished, Fitz came back into the room and put the blanket over her once more. Now all they could do was wait until the plates were developed to determine how to set the arm.

It was then that Dr. Montoya arrived and got his first glimpse of the quiet little ape. She showed no sign of distress, lying quietly in the midst of the crowd. Soon the developed X-ray pictures arrived, and the other doctors drew aside to examine them. Dr. Montoya turned back to Patty; she still gave no response. He touched her, manipulating the site of the fracture. He immobilized the arm to keep her from moving it until it was actually set. Even then Patty did not respond.

Once more the press poured into the room, this time to look at the X rays. They were excited and concerned, talking, filming, questioning the doctors.

Suddenly the door swung open, and in swept a tall, gray-haired, dignified woman, a full-length mink coat wrapped luxuriously around her, impressive in the sterile atmosphere of the hospital. Mrs. Ruth Oliver, public relations director for the hospital, had heard the commotion as she was getting ready to leave for the day.

"Please!" Her voice rang out imploringly above the din. "Please! This is a hospital! I'm afraid there's simply too much noise. Now I know you're excited, but there are sick people here, up and down these halls. You'll have to be quiet."

The press was subdued now, and when Patty was once again gathered up and carried down the halls to the animal care unit, they followed more calmly, speaking in low, restrained voices. A procession of people and paraphernalia streamed after Fitz and Ed in the immaculate hallways of the hospital.

The treatment room was far larger than the X-ray room, and the people did not press in over the table where the baby gorilla now lay. Instead, they wandered from doctor to doctor, interviewing them, or they prepared their equipment. The reporters studied their notes or reread speeches they had quickly jotted down.

Primarily, however, they were waiting for Dr. Michele, the orthopedic surgeon who was to set Patty's arm.

Dr. Arthur Michele happened to be in the hospital when Patty was brought in. He had stayed late that night and was thinking about going home for dinner when he heard himself paged on the intercom. He called Dr. Montoya.

"Listen, we have a little ape down here with a broken humerus. Can you come down and look at the X rays?"

"You've got a what? Listen, it's a little late for jokes. I'm going home for dinner in twenty minutes."

"No, really!" Dr. Montoya insisted. "Ed Garner brought in a gorilla. You know, Patty Cake. From the Central Park Zoo. The problem is, someone has to treat her."

"I'll be right there."

Patty looked up at Dr. Michele as he approached the edge of the table; as he reached toward her, she cringed. As he began to touch her, she twisted her body away from him and put her hand over her arm where it was broken. She looked up at him fearfully but never struck out or tried to bite. She bent her head down over the break and gripped it with her hand. Gently he turned her so that he could see her arm, and she made no attempt to fight him. Docile and obedient, she watched his movements from the corner of her eye and tightened her grip over the break. He had intended to stretch out the arm in order to examine it clinically. But when he saw her crouch in fear and watched her react to the obvious pain, he decided not to disturb her any further. The break was clear on the X rays.

The doctors stepped aside to discuss the treatment privately. Fitz stayed with Patty.

The doctors debated the various ways of setting Patty's arm. It was finally decided that they would use a combination of an arm and body cast, at least until the arm had begun to heal. It would hold the arm as still as possible and would prevent Patty from using it in any way. The fracture was sharply diagonal, and the doctors wanted to be sure that the two parts of the bone would not slip apart again once the arm was set.

There was a mild scramble in the room as the doctors disbanded and prepared for work. The press, who had been wandering about aimlessly, suddenly set up their cameras and focused them on the little ape.

The room quieted down when the doctors, bent over Patty Cake, began to work, speaking only to inform the press of the procedure and to help each other.

Patty looked terrified now. Her body was pinned to the table so that the job could be done as quickly and easily as possible. She was helpless, and terrified, held down by so many strange men she began to cry. Her plaintive screeches contrasted with the calm murmur of the doctors' voices. The lights of the camera crews were bright, brighter even than the sharp light of the examining room. Cameras whirred somewhere in the backround. Forms moved over her and touched her, held her, manipulated her. She opened her mouth wide, as if she were trying to scream. For twenty minutes this continued. When it was finally over, she was encased in a white plaster shell. Only her fingers showed through the heavy cast.

The press surged in around her again. Dr. Montoya picked her up and explained the procedure again in detail. When he finished, someone else reached out for her. Somehow, in the confusion, a young anthropology student, a girlfriend of one of the staff, managed to get Patty and posed. The cameras snapped and clicked. Once more voices rose and fell around the little ape. Again she began to cry. She opened her mouth in a pitiful grimace, drawing her mouth back over her teeth. A high *Eeeech! Eeeech Eh! Eeeech!* escaped weakly from her.

Finally, it was too much for Ed. He took her from the anthropology student and gently laid her on the table again. He would not allow anyone else to pick her up. One by one people began to drift away. By eight-thirty there were only a few stragglers left. Only Sharon Fama, who was working on a documentary film on veterinary medicine, was still shooting. Fitz leaned quietly against a wall, keeping an eye on Patty Cake. Dr. Michele and Ed were talking.

"What are you going to do now, Ed? You can't take her back to the zoo. She needs at least twenty-four hours of postoperative care. The keepers there wouldn't know what to do if something went wrong."

161

"Don't worry about it," Ed reassured him. "We'll figure out something."

"All right. But let me know if you need me. See you in the morning." Dr. Michele finally went home to his dinner.

For the last time that day Ed wrapped Patty Cake in the green and black plaid blanket and followed as Sharon carried her to Ed's office. It was here that Patty would spend her first night away from her mother.

As soon as he thought that Patty was calmer, Ed went out to try to find some kind of makeshift bed for her to sleep in. Sharon and Patty were left alone.

Sharon held Patty in her arms, but the little gorilla began to struggle. The anesthetic was wearing off; she was in a strange place, again, with a strange person. She twisted her head and tried to bite. She whimpered, *Eh, eh, eh, eh, eh,* and feebly clutched at Sharon's sleeve with her good hand. The young woman, who had never before held a gorilla, tried to soothe her. She did everything she could think of to help the frightened animal. She petted her and murmured to her and tried to nuzzle her. She rocked her back and forth and even straddled the baby over her hip and swayed back and forth gently as the thought a mother gorilla might do. But when Ed returned, Patty was still whimpering piteously.

Now Ed stayed with her, trying in vain to comfort her, while Sharon went to find something for her to eat. When she returned with a baby bottle filled with milk, she was accompanied by three nurses, whose payment for the food was a look at Patty Cake.

In the small office Sharon and Ed sat up with Patty Cake. They tried to feed her, cooing and murmuring at her as they offered her some food. She bit at the spoon and tried to twist away from it, from them. Baby food dribbled and splattered over everything. They looked at each other in resignation and put the food aside.

Patty reached up toward them, then pushed them away as her hands brushed against their clothes or their skin. It was all so strange to her, so new. But suddenly, as Ed bent over her, there was a difference. The awful, frightened grimace disappeared; she stopped crying. Her fingers played curiously with his beard, and for the first time she did not push him away. He held her close and could feel her muscles begin to relax. Her fingers clasped his beard tightly. She put her head down on his chest.

By nine thirty Patty was calm. Ed laid her in the little carton he had found and covered her with the plaid blanket. He lifted it onto his secretary's desk in the small room next to his office. Only when he saw she was asleep did he call his wife to tell her that he would not be coming home that night.

At ten o'clock Sharon went home. Ed was alone.

He stretched out on the couch in his office but could not sleep, afraid that he would not be roused if Patty woke and cried. Sleepily he pushed himself up from the couch and went to look at her. Her left hand was clutching the blanket as she slept. She looked vulnerable and small in the half darkness.

Carefully he picked up the box and carried it into his office. He set it down beside the couch and then lay down. He felt better, having her closer to him. He, too, drifted off to sleep.

But he slept lightly, afraid that she might wake and need him. When she did awaken, crying, he gave her some of the milk from the baby bottle. She soon fell back to sleep.

All that night the door would open, and someone would tiptoe in to look at the sleeping gorilla. Ed would be startled out of his doze to see a dark shadow standing there.

"I've never seen a baby gorilla before," the shadow would whisper. "Someone told me Patty Cake was here." It would stare down at the sleeping infant for a moment and then softly back out.

Toward dawn Ed heard a little cooing sound, and then a cry as Patty wakened in this strange new world. He looked down at her and picked her up. He sat comforting her until once more she fell asleep. His head, too, sank back against the sofa, and he napped, waking now and again to look down at the infant in his arms.

When Dr. Michele arrived the next morning, that was the way he found them: Patty Cake fast asleep, nestled in Ed's arms.

CHAPTER 24
Grieving
March 21, 1973, at the Central Park Zoo

March 21 was a dismal day at the Central Park Zoo. Patty was gone. I knew the cage would seem empty without her. Unenthusiastically I headed for the Lion House.

"Hi, kids!" I tried to keep my voice as cheerful as usual, almost singing the words.

Lulu and Kongo were still separated. In the number three cage Lulu lay motionless on the crossbar. As I spoke, she turned her head away from me.

Kongo glared at me; his eyes were hard and furious, and my heart sank as I saw him. He had never looked at me that way before. He stalked along the length of the cage, continually going to look in on Lulu next door, then striding back again. Violently he charged anyone who came near. A defensive wariness drove him back and forth across that cage, scanning the surrounding area with fixed and furious intensity. He never looked into my eyes.

Richie, squatting next to the bars of her cage, murmured softly to Lulu. But she would not turn her head toward him. She lay still, absolutely quiet, a counterpoint to the violent fury of her mate next door.

Suddenly a balloon broke! The sound burst into the air.

Lulu bounded from the crossbar. She moaned once and backed

into a corner, pressing herself against the wall. Her terrified eyes swept the house. She waited, cowering. Nothing happened. There was no shot, no dart to pierce her leg. Timorously she crept forward, looking for the source of the noise. Richie called her softly, and eventually, nervously, she sat at the bars in front of him. Her hands gripped the bars tightly; her head swiveled sharply at any sound. Richie murmured sweetly at her, and she looked into his eyes. But when he slowly reached out his hand to pet her, she sprang back convulsively. Her eyes were wide with fear.

She sat in the middle of the cage, her head jerking from side to side. But in a few minutes her energy seemed to abandon her. Her body lost its rigidity, and she sank into listlessness again. Wearily she climbed back onto the crossbar and lay down.

Her breasts were swollen with unneeded milk. Every once in a while her fingers slowly went to them, pressing them in a heavy, indifferent way, a weary and preoccupied movement. She barely responded to us at all; she averted her eyes when I spoke to her and did not even stir when I offered her a lollipop. She did come down from the crossbar when Luis came by with the hose. Halfheartedly she plodded to the bars, drank a bit from the dripping hose, and returned to her perch without ever looking at any of us. She remained there, staring into space. An hour passed without any sign of movement. I ached, wishing that she knew that her baby still existed in another place. But this cage was Lulu's entire world.

I gave a lollipop to Luis, and eventually he coaxed her into taking it from his hand. She merely held it limply, staring at it, slowly twirling it with her fingers. Then even that little movement stopped, and she was still again. The lollipop fell to the floor. A quarter of an hour later Luis handed it to her again. She put it into her mouth, but she did not bite into it. She did not even suck on it, and it soon slipped out of her mouth and fell again. It broke; shiny slivers of crystalline candy showered across the floor.

I moved over to Kongo's cage. Although he swiped his lollipop out of my hand with unprecedented violence, he did not look into my eyes either. He turned immediately and strode deep into his cage. When he looked toward me again, his eyes brushed over

166

mine as if I were not there. He went to the bars again and stood looking at Lulu.

When Luis opened the door between the two gorillas, Lulu stirred, suddenly alert. She dashed into the cage, glanced briefly at Kongo, then rushed purposefully past him. She trotted around the cage, racing from corner to corner. Her eyes swept the bars and the walls, probed the corners of the cage, the base of the pole. She climbed to the windows and searched the little iron ledge beneath them. She peered outside, straining to see down into the cages there, then swept down to the floor and bent to squint through the peephole in the door. Again she took a turn around the cage, this time with less enthusiasm, less vitality. Was it with less hope? She went to the door and bent down to the peephole.

There was a sudden change in the movement of her body. Slowly she straightened herself and paused, drifting to the very center of the cage. She simply stood there once again, her head shaking slowly from side to side, still searching, until even the slightest movement ceased. I knew that she had been looking for her daughter.

Kongo had been watching her from the side of the cage. When she stood dejectedly still, he came to her. He lifted his leg over her rump and tried to mount her. She staggered back, and his leg slipped to the floor. Again he came up behind her, and this time she deliberately shook him off. There was no annoyance in the movement; it was, rather, a weary, apathetic gesture.

He approached her again and again, gently trying to mount her, as if to say, "I'm here." But the reaction that he was looking for never came. When he came around her to look into her eyes, she did not even turn her head toward him.

He gave up.

Listlessly Lulu returned to the peephole. He followed her there and then around the cage as she moved around and around, looking into the same corners, scanning the same bars time after time. His eyes never left her.

Kongo seemed intent only on Lulu. He never did search the cage for his missing infant.

Yet when the keeper, Veronica Nelson brought the baby baboon

167

Gertrude, over to the cages, Kongo saw her and sped to the bars. He stared at the scrawny red-haired monkey, and Lulu rushed up to join him. She put her hand on Kongo's arm and narrowed her eyes in intense concentration. But then her hand dropped, and her body lost its poised, rapt expression. She turned again and moved off to lie down in a distant corner.

She often seemed to feel the baby's presence on her body. Her arm reached out in a familiar, cozy gesture as if to enfold a child which lay nestled on her stomach, as if to put her arm around her and slide her up to hold her against her breast. But there was only air, and the gesture went unfinished. The gesture was so small, so familiar, automatic, and incomplete that it fairly screamed that the baby was no longer there.

Lulu lay staring into space and gingerly touched her engorged breasts once again. Her milk was dripping. With bleak, helpless eyes Lulu went to Kongo and put her hand on his back. She stayed with him then. They sat together, motionlessly, his hand on her crest, without moving for a long time.

Kongo finally left Lulu to gather some fruit, and when he returned, he sat beside her and pressed his penis into the soft hair on her leg. Swiftly she rose and presented herself to him. He examined her carefully and poked his finger into her vagina. With sudden quickness he pulled her rear toward him and began a copulating movement. He tried to pull her down onto him, but she was unyielding. With a quick, impatient movement she flung him off. She ran. When he did not pursue her, she sank to the floor and slumped there.

Again there was a long period of depressing stillness.

Without warning, Lulu sprang from the floor and flung herself toward Kongo. She hurled herself at him, and he backed into a corner, raising his arms up across his chest and face as if to protect himself from a sudden assault. But she only stood leaning heavily against him. She looked up into his face and moaned softly, hooting dolefully at him. She seemed to be searching his face for something. But he only stood motionlessly, looking down into her eyes. In a moment she seemed to lose her energy. She dropped down onto all fours; he backed away; she looked after him. Again despondency overtook her.

168

Gently Kongo touched the cuts and scrapes over her eyes. Milk dripped from her nipples; Kongo's hand went to her breasts. She ran her fingers over his hand as he drew out the milk, then licked it off his fingers. Drops of milk trickled like tears down her dark breasts. They left paths that crept into the hair on her belly, then disappeared. She tapped her distended breasts and squeezed them gingerly, pressing them, and pulled the milk from her nipples. How painful it must have been by now.

She put her hand on Kongo's head as he touched her, and she seemed to relax a little. He began to lick her ear where there was still another crimson gash. He seemed to be sucking on her cuts, one after another. And she settled down. He stroked the hair on her neck, and she lay quietly as if his touches were soothing her. He watched her with concern, and soon her eyes, which had held only the wide, vacant stares of uncomprehending grief, slowly closed. She fell asleep. Only then did Kongo close his eyes, as if he knew that now he could relax his watch.

Lights! A sudden violence of light flooded the cage with power and unrelenting brilliance. Cameras whirred. They were in the keeper's walk and in the keeper's room.

"Get that shot!"

"That's a good one!"

Voices, loud and urgent, intruded into the privacy.

The television crew crowded into the small space in the keeper's room and flowed into the walk itself. They searched, with penetrating unconcern, into the privacy of the animals' helplessness, made harsh and public in the glare.

Lulu sprang up as the lights flashed. She stood bewildered, her eyes once more red with fear. She backed away toward the dark and safe private, hidden cage. But Kongo would not leave the home cage, and she would not leave him.

Kongo, furious, backed up and then charged the trespassing men. He lashed his hand toward them between the bars.

And they laughed!

Helpless in his fury, Kongo charged again. The cameramen winced at his power. But the camera kept whirring on.

The lights bombarded the gorillas. Lulu stood transfixed and frantic. Kongo's eyes were marble-hard, like two black agonized stars. He swept up an orange and flung it at the men. It bounced off the bars and rolled back to his feet.

The keepers implored the men to leave. But they wouldn't, not until they were finished. At last the lights went out, and the threatening sounds of the camera ceased. The men lugged their equipment out of the aisle and out of the keeper's room, laughing and joking as they packed up their gear. As they passed me, one of the men whispered, "Don't tell anyone! Don't let Fitz know."

Kongo stood stock-still, watching them, until the last man had left. Then it was quiet and dark once more.

Lulu quieted, and Kongo went to her and she to him. He put his hand across her back. The soft light once more enveloped them, bringing them back to themselves. Together they moved to the middle of the cage and lay quietly on the floor, barely moving. They lay awake for more than an hour, Kongo's hand touching her, his eyes watchful.

That is how I left them that night—Kongo lying next to Lulu, comforting her; their heads together, their hands entwined.

CHAPTER 25
Patty Leaves Us
March 22, 1973

Fitz and I sat on the black naugahyde sofa in Ed Garner's office. At his desk Ed twisted a pencil idly between his fingers. We had long ago run out of conversation.

Ed stirred. "Well, I might as well try again." He picked up the phone and dialed the number of the Bronx Zoo.

"Hello. Is Dr. Dolensek there?" There was a long pause. For the thirtieth time I stared at some photographs of mating rhinoceroses that hung on the wall near Ed's desk.

"Hello?" He listened for a moment. "Well, please tell him that Dr. Garner called again. We've been trying to reach him for hours. Do you know when he'll be back?" There was another pause. "OK." He sounded disheartened. "Thank you."

He hung up and looked at us helplessly.

"Same story?" I asked.

He nodded.

We had been sitting there since noon. It was now four o'clock.

Yesterday Patty had been taken to the Bronx Zoo.

At the time it had seemed the sensible thing to do. The Bronx Zoo had a complete veterinary hospital on its premise, and a complete medical staff, which not only included a fine veterinarian em-

171

ployed solely by the zoo, but various trained technicians and assistants and a long list of consulting physicians and veterinarians as well. The Central Park Zoo had only the primitive facilities of some empty cages and a small caged-off room behind Fitz's small office. Ed was the only doctor readily available to the Central Park Zoo, and he had obligations to fulfill elsewhere. He could not drop everything to sit with Patty around the clock. And it would not be possible for Patty to remain at the hospital indefinitely.

Fitz and Ed were very concerned. They discussed the various possibilities open to them and finally concluded that the Bronx Zoo had all the facilities necessary for Patty's best possible care, at least until her initial special needs were satisfied. They called William Conway, director of the Bronx Zoo, and asked for his help. Eventually Mr. Conway, Brad House, the curator of mammals, and Dr. Emil Dolensek, the veterinarian, agreed to take Patty on an interim basis.

When Fitz arrived at Flower and Fifth Avenue Hospitals to pick up Patty Cake, he found the press already gathered outside. They were also inside, taking pictures and conducting interviews in Ed's office. When Fitz finally picked up the still-groggy, frightened gorilla and carried her through the halls to his car, they trooped along behind him. Some of the newspeople asked if they could ride with him to the Bronx Zoo. He declined. When they arrived at the zoo, more press was waiting for them. But there it stopped. For the administration of the Bronx Zoo did not allow the press to enter its grounds. It is a private enterprise and could readily dictate its own terms. Fitz drove through the gate and past the throng of reporters.

Dr. Dolensek was waiting. When Fitz and Patty arrived, he led them down a long hall to an isolation room. This was to be Patty's nursery.

Fitz laid her on the table and for the last time unwrapped the blanket that had traveled with her and comforted her during these last two difficult days. Then he stepped back, and Dr. Dolensek came forward to take charge. He looked her over quickly, ordered various blood tests, examinations of urine and stool, a polio vaccine.

A sweet-looking woman came forward to look at Patty's frightened face. She was Caroline Atkinson, who would be Patty's surrogate mother at the Bronx Zoo. Mrs. Atkinson's cot was already standing next to a wall. She would eat there, sleep there, be there. She bent over the little ape and smiled. Patty looked up into kind, sympathetic eyes.

When the examination was over, Mrs. Atkinson carried Patty to a playpen in the middle of the room. Gently she laid her inside and then sat down in a chair next to it, murmuring comfortingly all the time. Patty seemed small and alone as she lay in the playpen. The plaster cast looked heavy and bulky on her body. Mrs. Atkinson covered her with a blanket, and soon Patty's eyes closed.

Fitz waited until Patty had drifted off to sleep. Then he turned and walked away.

Fitz called the Bronx Zoo for bulletins on Patty's progress at regular intervals. The following morning I happened to be in his office. His response to Dr. Dolensek brought me upright in my chair.

"What?" he said sharply. "How much?"

A look of intense concern crossed his face. He took off his glasses and rubbed his eyes. "Well, let me call Dr. Garner and see what he says. I'll get right back to you."

"He says that her hand is cold," he explained anxiously, "and the fingers are a little swollen. He says that the cast is too heavy and too tight. It should come off. . . . Hello, Ed?" He repeated what he had just told me.

"I'd like to see it," Ed said at the other end of the line. "I'll get hold of the orthopedic man, and you can meet us here. We'll all go up together and take a look."

"Sure." Fitz looked at me. There must have been pleading in my eyes, for he said, "Look, Ed, Susan is sitting here. I think she'd like to know what's going on."

"Bring her along. Glad to have her."

"Fine" Fitz said, "I'll call the Bronx and let them know we're on our way."

But Fitz could not get through to Dr. Dolensek. He did speak with someone but could not get any information. No, Dolensek hadn't left any messages. No, he had left the hospital. No, they did not know where he had gone. No, he hadn't said when he'd be back.

Fitz shrugged in resignation and turned to some papers on his

173

desk. "I'll try again in a few minutes," he said. "In the meantime, let's get some work done."

I wandered over to the Lion House to spend some time with Lulu and Kongo. By the time Fitz and I left for Ed's office an hour later we had still heard nothing. Dr. Dolensek had still not been located.

At twelve thirty we were in Ed's office as he called Dr. Michele. He listened intently and then reported back to us. "He says that the swelling shouldn't mean too much right now. It's really the first twenty-four hours that are crucial. We've passed that now."

But we were still somewhat uneasy, for, of course, we really did not know what Patty's arm looked like or how much swelling had taken place. If her circulation were cut off, it could be dangerous. But there was nothing that Fitz or Ed could do without seeing her. And they did not feel that they could simply arrive at the Bronx Zoo uninvited.

"I'm beginning to wish that we'd kept Patty with us," Ed said. "This is like living in limbo."

We took out sandwiches, ordered danish and coffee, and settled down to lunch. Another hour passed, and still no word. Dr. Garner called the Bronx Zoo again. Someone said that Mr. Conway and Mr. House had left town that morning for some meetings somewhere. They would not be back for a couple of days. No one could get in touch with them. James Doherty, assistant curator of mammals, could not be found.

It was becoming discouraging.

"Look," Ed said, "it can't be too bad or they'd be on the phone with us right now. She's too precious for anyone to be playing around with her. They know that."

Fitz nodded. "Yeah," he said in a flat monotone, "sure."

"Let me try again." For the fifth time in two hours Ed picked up the phone and dialed. No one could be found.

Out of frustration we began to talk about simply going up to the Bronx Zoo, walking in, and taking Patty home.

"But I don't think it would be right to barge in there," Ed said. "We can't do that."

"I guess not. After all, we asked them to help."

174

"Yes, but I didn't think it would be like this. We don't know what the hell is going on with her."

At three o'clock the telephone rang. We jumped. It was only Central Park calling Fitz. Ed handed him the phone and we saw his face turn grave as he listened.

"OK, I'll be right down."

"I've got to get back to the zoo," he told us. You know those lesser pandas we just got? The female died. She's been listless since she got here. I'll pick her up and bring her back for a necropsy. At least we can get that done."

"Just what he needed!" Ed said as Fitz disappeared behind the door. "That panda arrived sick. Boy, what a lousy couple of days these are."

Time passed slowly, broken only by fruitless attempts to reach Dr. Dolensek. An hour passed before Fitz reappeared with a weighted burlap sack in his hand.

"Anything here?" he asked.

"No. Let's get that necropsy done. Susan'll call us if they call."

"Sure," I said, "you go ahead. Just tell me which lab you're using. Just in case. . . ."

But I should not have been concerned. There was no phone call.

Finally, Ed spoke with Jim Doherty. But Doherty could give him no information. No, he wasn't sure where Dr. Dolensek was. No, he couldn't take the responsibility for anything in connection with Patty Cake. He was sure that Dr. Garner would understand that. He would leave a message.

It was 6 P.M. I found myself staring at the books behind Ed's desk, reading off the titles to myself. *Parasitism in the Lower Vertebrates, Anatomy of the Dog.* My glance moved over to the rhinos once more. They were still mating.

The phone on Ed's desk jangled sharply. His hand streaked out to it.

This time it was Dr. Dolensek.

Finally!

He was very serious and very assured. It had been a long day for him, too. He had been unable to call before, he said, because he had been in consultation with other doctors all day. They had been discussing Patty Cake—her test results, X rays, general health. The results were not very encouraging.

As he listened, Ed's face became more and more downcast. According to the consulting physicians and Dr. Dolensek, Patty was in very bad condition. It was not only her arm that was a cause of concern, but malnutrition, parasites, depression, and general bad health. As far as her arm was concerned, they said, the cast was too heavy and too bulky. It was apparently cutting off her circulation.

"How much swelling is there?" Ed asked. "How cold is the hand?"

It was minimal, Dr. Dolensek reassured him. But the consensus was that the cast should be removed as soon as possible. From the subsequent X rays there seemed to be a bone separation of about one-half inch, probably caused by the weight of the cast on Patty's arm. On top of everything else, since the cast forced her to lie on her back, Dr. Dolensek feared that she would easily contract pneumonia.

We were shocked. We had expected that any examination would disclose the presence of worms. Most zoo animals have them and are dewormed on a regular basis. We had also expected her weight to be less than that of an animal raised on formula and vitamins. And we knew that Patty was small. But actual malnutrition? We could not fathom it. She had always seemed so strong. It was logical that she was depressed. Any animal taken from its mother under these conditions would show strain. But to hear these complaints listed in a long and relentless statement was heart-rending. According to Dr. Dolensek, those responsible for Patty Cake had done everything wrong. I simply could not believe it.

The telephone interrupted our depressing conversation. This time it was Ted Mastroianni, deputy commissioner of the Parks Department. Apparently Dr. Dolensek had already been in touch with him and had discussed Patty Cake's condition. Mastroianni was very upset and concerned by the frightening report. His first reaction, upon hearing Dr. Dolensek's list of complaints, was that Patty Cake be immediately brought back to Central Park. Then he and Commissioner Clurman thought that perhaps Patty would receive better care at the Bronx. Perhaps it would actually be dangerous to the animal to bring her home. Aside from all the ailments mentioned, Dr. Dolensek's primary argument for keeping her at the Bronx Zoo concerned Patty's mental state. She was apparently in a state of shock, and another physical change for her at this time could mean only further trauma. She should be left where she was to avoid additional handling. Dr. Dolensek felt that there was real danger for her, perhaps the danger of death.

Mr. Mastroianni and Commissioner Clurman wanted to be assured that Patty would receive the best care. They had listened with grave attention to everything Dr. Dolensek had said. Now they wanted Ed's opinion.

What could he say? He, too, wanted what was best for Patty. He had to agree that if she was in such a depressed state, it was important to alleviate any further stress or anxiety which might occur if she were moved. It was essential for her to be quiet and undisturbed.

The calls went on and on. Dr. Dolensek told Mr. Mastroianni that he must have complete control over the animal and the responsibility for any decisions to be made. Unless he was given that control, he said, the proper care which the Bronx Zoo could provide for Patty could not be given in full.

Mr. Mastroianni replied that the main objective was to take care of Patty Cake. But he also felt that any decisions regarding her health should be made with the consultation with Dr. Garner.

Dr. Dolensek was adamant. He refused. "I can't go through this again," he said.

Again the indirect phone calls went back and forth. Finally, a compromise was reached. Patty Cake would remain at the Bronx for an undetermined amount of time. If Dr. Dolensek felt it necessary, the cast would be removed and X rays taken. Any medication that would be used would be given with the agreement of Dr. Garner and with the knowledge of Deputy Commissioner Mastroianni. Commissioner Clurman and Dr. Garner would be continually informed of Patty's progress and would have direct access to her, as would Mr. FitzGerald. Ed then asked if I would be permitted to continue my work with Patty while she was at the Bronx Zoo. There were two reasons for this. First of all, we felt that it would

177

be important to record her emotional responses to the radical changes in her life. Second, I would be a link between her new life at the Bronx Zoo and the one she had known at Central Park. It would make the transition back to Central Park easier for her. It was hoped that her keepers would be able to see her for the same reason.

Dr. Dolensek made it clear that for an unspecified period Patty would not be allowed to associate with Mopey and Hodari, the two gorilla infants that lived at the Bronx Zoo. She would remain in isolation at the hospital until she had received all her shots, had been dewormed completely, and was considered clean enough not to endanger Mopey and Hodari with any disease she might have been exposed to at home. Her weight would be checked and recorded several times a day. She would have the constant care and companionship of Mrs. Atkinson, as well as any special attention that might be required. She would be well taken care of there.

It was only after the arrangements were completed that Dr. Dolensek and Ed Garner finally spoke directly to each other. When Ed finally returned the receiver to its cradle, we stared bleakly at each other.

"Well, that's that," he said dejectedly. He dialed the number of the orthopedic surgeon, who had been waiting all day, and told him, "Go home. It's all over. Sorry you had to wait."

It was nearly 9 P.M.

Outside, the night air was cool, and I breathed deeply. The breeze seemed to stir us from our gloomy lethargies. We formed a sad little circle under the light of a streetlamp. Fitz was the first to speak. He looked worn. "Well," he said, "thanks for everything." We shifted around for a moment. There was really nothing much we could say.

"Anytime," Ed said.

"Night," Fitz said, and turned away. He walked down the street, his hands deep in his pockets, his shoulders slouched.

Ed and I shook our heads as we watched him. "What do you suppose'll happen now?" I asked.

"I don't know." He smiled weakly. "They'll take good care of her," he said. "They've got everything over there."

"I know."

CHAPTER 26
Valentine's Day at the Bronx Zoo
Mopey and Hodari
February 14, 1973

Yes, they had just about everything. I had had a small glimpse of it about a month before, when I had been to the Bronx Zoo to see its two gorilla babies: Hodari, who was born there, and his playmate on loan from the National Zoo in Washington, D.C., a gorilla with the nearly unpronounceable name of Mgani-Mopaya, or, as he was called, Mopey.

Toward the beginning of February, Brad House, then curator of mammals at the Bronx Zoo, had come to Central Park for another brief visit and a glance at Patty Cake. As luck would have it, she lay fast asleep on the floor in the center of the cage. She looked her absolute tiniest as she lay on her stomach, her arms outstretched over her head. Pretending to work and straining to listen, I caught the words "Our Hodari is twice as big and a month younger."

That was all I heard before he and Fitz moved on.

I was very disappointed, for I wanted Brad to see Patty at her best: crawling, eating, exploring, climbing. And there she lay snoozing, dead to the world.

Now I was anxious to see the Bronx Zoo babies. I had heard of these two gorillas previously. The administration of the Bronx Zoo had thought that Hodari would be the first gorilla born in the New

York area and had begun publicizing his impending birth during the summer. But then Patty was born in September. When Hodari made an appearance the following month, he came into the world without fanfare. He was left with his mother, Sukari, until he was thirteen days old. Then Brad and the keepers saw Sukari dragging Hodari across the floor. She had not allowed her infant to nurse that day, and the men became more and more anxious for his safety, until they feared for his life and took him from her. After a stay at the veterinary hospital on the premises, he was taken to the home of the curator of reptiles, Wayne King. King's wife, Sherry, was to hand-raise the baby gorilla.

But a hand-raised gorilla usually has few, if any, of the experiences which will later allow him to relate normally to other gorillas. The Bronx Zoo did not want Hodari to be in this position, to be raised alone, and began to look for another little gorilla to be raised with him.

Mopey had been born in the National Zoo in Washington, D.C., five months before Hodari. He had been taken from his mother within an hour after his birth and had been hand-raised until December, when he arrived at the Bronx Zoo and was introduced to Hodari. They were now being raised together, and I wanted to see them.

I called Brad House to ask permission. To my surprise and relief, he remembered me and said that I was welcome.

"The two boys, as we call them," he said, "are being taken care of by Mrs. King. Just call her to make an appointment and then let me know."

Elated, I called Sherry King.

"Sure, anytime," she said. "As long as you don't have a cold or anything. You know how susceptible they are." I assured her that I was perfectly well, and we made the date for February 14, Valentine's Day. I would meet Brad at the administration building, and he would take me to Sherry's to meet her and the gorillas.

February 14 was a beautiful, sunny day. I hadn't been to the Bronx Zoo since I was a child. Now, as the bus wound slowly through unfamiliar streets of the Bronx, I tried to recall what it was like. Huge, I remembered, and green, with herds of animals in

180

large enclosures, spacious with the illusion of natural habitats. I remembered a maze of paths that led through a park setting to open up here and there to exhibits of all sorts of wonderful and exotic animals. It had the reputation of being one of the finest zoos in the country.

I had brought some of the drawings I had made at Central Park, hoping they would show that I was serious about the animals and that I had some knowledge of them.

The ride was taking far longer than I had expected. Would I be late? That was all I needed, to be late! Sure enough, by the time I saw the park before me it was nine o'clock. When the bus door opened, I grabbed my twenty-pound sack of equipment, flung it over my shoulder, and leaped from the bus. I ran lopsidedly and awkwardly down the street to the main entrance of the zoo.

The gates were closed.

There was no one there.

I looked down the road for any sign of another entrance nearby. But there was nothing in sight—only the tall, carved gates that loomed in front of me, impenetrable. Peering through an open space above the back of a sculptured tortoise, I could see cars moving at the end of a long driveway. Miraculously one of them started toward me. I began to call out at it, frantically waving a piece of white paper. Just as the car began to turn, the driver saw me. The car stopped. With nerve-wracking slowness the man got out of the car and ambled toward me. "Can I help you, miss?"

Slightly hysterical, I explained.

He stared at me thoughtfully for a moment, then seemed to make up his mind.

It was very reassuring, at first, when he began searching for a key to let me in, but when nothing turned up and time was very quickly passing, I began to feel panicky again. We were staring helplessly at each other through the large space above the tortoise when I had an idea.

"Do you think I can fit through there?"

"Gee, lady, I don't know."

But I was already taking off my winter coat. He looked at me incredulously as I passed it through to him. In another moment my equipment and my shoes were on the other side of the gate. I felt very alone standing there, shivering, on the empty street as the traffic swished by on the highway behind me. What if I couldn't get through? I glanced at my watch. Nine fifteen. I climbed onto the tortoise's back.

After one false start I just managed to squeeze through and slid onto the shoulders of the man on the other side. I was in! He threw the coat over my shoulders as I scrambled to put on my shoes.

"It's a good thing you stayed thin," he called as we raced for the car. "You coulda got stuck!"

In a few moments I was breathlessly running up the steps to the administration building, leaving a trail of bobby pins on the stairs behind me. My coat was dragging on the floor, my hair loose and streaming, the heavy sack banging against my legs as I flung open the door and rushed inside.

"Mr. House is not quite ready," the receptionist said. "If you'll just take a seat. . . ."

By the time Brad appeared I was once more neat and calm. My hair was up in its knot on top of my head. I had put on my coat, and my breathing was normal. I was, I thought, the picture of professionalism.

He nodded hello at me as he came by. "Ready?" he said. I nodded and followed him as he sped down the stairs. He led me to a golf cart parked in front of the building. Two beautiful huskies, sitting impatiently in the back of the cart, welcomed us with loud barks of excitement.

"This time of day they usually run loose," Brad explained. "They're all excited." He turned to the dogs. "Quiet down now!" he ordered. Brad helped me into the front seat, handed my equipment to me, and we were off. The cold air rushed past us as we drove down paths that twisted through the park woods. Every once in a while the huskies poked me from behind and I could feel a wet nose on my neck. Around us the tops of the bare trees looked like soft brown feathers in the sunlight. The day was beautiful.

Brad and I really became acquainted as we drove past a herd of buffalo and then up a winding path surrounded by wintry countryside. I relaxed. I enjoyed my ride up to the "Farm," where the Kings and Brad lived. The city that surrounded us seemed far away, and all that I saw was land, trees and sky. Only the swishing sound of the highway traffic nearby reminded me of the tar and gray cement that were just a few yards from us.

We turned into the grounds of the "Farm," and Brad got out to open the gate. This once was a real display farm with sheep, cows, and other livestock for city kids to see. But now, although the "farmhouses" and grounds remained, the farm no longer existed. My heart thumped with excitement as we parked the golf cart in a driveway next to a beautiful big house. Brad put the huskies in the backyard, let me into the house, and we climbed a narrow flight of stairs. As he rang the doorbell, we could hear dogs barking inside. In a moment the door opened a little more than a crack, and a young woman looked out at us. She held a baby gorilla in her arms; two big, graceful dogs swarmed around her legs and poked their noses out at us.

"Wait a second," she said. "Let me get the dogs downstairs." She closed the door. When it opened again, the dogs were no longer there, and she stood welcoming us with the baby gorilla clinging to her.

Hodari looked at me with great, solemn, timid eyes. Patty's features were so delicate that Hodari's head looked large and square to me. He stared at me for an instant, then turned away from me, buried his face against Sherry King's shoulder and clutched her even more closely.

"He's a little shy," Brad explained as we went in. "We're hoping that he's coming out of it." He patted Hodari's head as we walked in. "How are you, fella?"

He's so big, I thought. I was awed by his size. I stumbled against a playpen that stood next to the door.

"Careful!" Brad put out his hand to help me. "There's stuff all over the place."

True. Cartons and toys were strewn over the floor, and the playpen, which seemed to be lined with sheets of hard, transparent plastic, was filled with toys and hung with mobiles. I put my things down and followed Sherry, making my way around the clutter on the floor.

It looked as if the place had been hit by a hurricane, and the hurricane's name was Mopey.

There was nothing timid about this animal. He was charging

183

across the room, sliding across a piece of masonite to crash into one of two or three large cartons. At the same time he reached out to hit at a woman sitting nearby on the floor. As I stepped into the room, he stopped in midaction and looked up. Sherry followed, Hodari still in her arms, watching Mopey with intent and anxious eyes.

High on his knuckles, his back arched, legs stiff, Mopey stared at me, boldly confident. He took two or three steps toward me, then stopped, still staring. He picked up a plastic block and threw it at me. As I dodged the flying block, he stopped and stared again. Satisfied then, he turned to his former play and charged a garbage can. But I saw that he was still watching me from the corners of his eyes and knew just where I was and what I was doing. He was the epitome of the male gorilla in miniature.

Sherry introduced me to her mother, Barbara Ryther, and we joined her on the floor. Trying to make myself as unobtrusive as possible, I settled back in a corner to watch. I took out my pencils and papers. Mopey was already sneaking around me, coming closer and closer as the minutes passed. When I took out my drawing pad, he stared, unabashed curiosity in his eyes. But he turned away to Brad and flung himself at him. He obviously knew and liked Brad House and went lunging at him, hitting out and pulling his jacket, his hair, whatever he could reach. I could see that Brad, too, loved this play. Crawling around on the floor with Mopey, he allowed himself to be engaged in a mock attack and then threw off the sturdy little gorilla, rolling him around on the floor, laughing as Mopey came charging back for more.

Hodari, still quiet, watched me timidly, shrinking against Sherry if he found me looking in his direction or if Mopey came too near him. He sat next to her on the floor and clung to her leg. When she shifted her position, he turned to her and clutched her arm.

"I'm not going anywhere," she said reassuringly. But he reached up to her pleadingly, and with a sigh, she gave in and picked him up. He twisted around in her lap in order to keep his eye on me or on Mopey, who was chasing Brad around the room like a powerful little whirlwind. I turned again to watch Hodari and Sherry when suddenly I realized that the room was unaccustomedly quiet. When I looked up, I realized that Brad had left quietly.

Now that he was gone, Mopey turned to little Hodari. Hodari

184

squeaked a little *Eeeh*! and sank deeper into Sherry's arms. Mopey advanced menacingly. "OK, Mopey," Sherry called and turned toward him just in time to catch his hurtling body. As Mopey struggled to climb up onto her head, she gasped, "He won't play as roughly with me. Somehow he's much rougher with the men."

With all the energy in his body, Mopey rushed her again. She pushed him away, and he fell over backward, chuckling. He scrambled up and came running again.

On the other side of the room Hodari began to relax, shyly he began to move around on the floor. Like Mopey, he walked on his knuckles. Sherry told me that neither of the "boys" had ever walked on the flats of their palms the way Patty did.

Hodari moved around more slowly than Patty usually did; he was far more unsure of himself. He crawled by using both arms and only one leg. I wondered if the leg that he pulled behind him was weaker than the other. He headed toward a blue wooden block he had set his sights on. And, as Patty would have done, he put it right into his mouth. He was gnawing on it, his round eyes still watching me, when suddenly Mopey swooped down on him from the rear, bowling him over. Hodari yelped. Sherry reached over to pull Mopey away from the smaller ape. Hodari was cowering and trying to reach his protector.

"Oh, no, you don't!" Sherry yelled as Mopey made another attempt to bully his smaller agemate, and she half tossed him in the opposite direction. Again he rushed at Hodari, she intercepted him halfway. She put the plastic garbage can over his head, and he lifted it off, laughing.

Again and again Mopey reached for Hodari, only to be swept away by one of the women. And each time Hodari shrank away from him and clung wide-eyed to Sherry, his surrogate mother.

Suddenly Mopey, the running tornado, disappeared behind me. As I turned, I felt a sudden, painful yank on my long hair. "Mopey!" I cried out sharply. The next thing I knew was that I felt a heavy body heave itself onto my back and begin to climb. He was reaching for the top of my head, and he pulled at it until my hair dripped into my eyes. Another groping hand crossed over my face from the other side and poked a finger into my eye.

"Mopey!" I yelled. Sherry tried to tear him off me. He was

stronger than I was, and it took both of us to free me. Once on the floor again, he headed for my paper; little gorilla handprints appeared on the white sheet. I flung him off, and he came charging back. Evidently he thought it was a terrific game.

"This goes on six hours a day," Sherry said. "Look." She held out her arm to show me the bruises there. "They're all over me," she said. "It's exhausting." I thought of Patty Cake being bounced and tumbled by her mother, laughing as she somersaulted happily in the air.

As Mopey became occupied on the other side of the room, Hodari now ventured out, bravely leaving his surrogate mother. He kept an eye on Sherry as he moved from one toy to another. In his wanderings he came closer to me and stopped. He watched me openly, glanced back at Sherry for reassurance, then came slowly and hesitatingly toward me. Tentatively he put out his hand and clumsily brushed my knee. He scooted backward quickly, his eyes glued to my face. When I did not move, he came toward me again, touched my paper, then turned away. He played happily by himself, wiggling across the floor, his nose to the ground as he explored, his diapered rear waving like a flag above him.

Suddenly he fell; his head clunked against a metal pole of the climbing bar.

"Oh!" Sherry and Barbara both rushed to him, picked him up, and examined his head. He clutched Sherry's arm.

"He looks OK," Sherry said, trying to disentangle herself from his grasping hands and squirming body. It was a few moments before he began to wander off by himself.

Soon Mopey struck again. This time when Sherry swooped him up into her arms, she carried him off, away from Hodari.

"I know what will keep you," she told him. "It's lunchtime, so you come with me." She carried him into the kitchen, and lifted him onto the sink. "Now you stay there, and don't move! He gets into everything," she told me. "He loves to pull out all the pots and pans if I don't keep track of him." Mopey was already reaching for the water tap. "Here." She handed him a bottle of juice, and he sat drinking it. He was quieter than I had yet seen him.

"In Washington," Sherry said, "he had a blue bottle. Here it's clear, and he's fascinated by watching the liquid slosh around in it. It'll keep him occupied for a few minutes anyway."

Soon the apes' lunch was ready: applesauce, Enfamil, and cereal. Barbara tied bibs around the gorillas' necks. "It's too much of a mess if we leave it off," she explained. "They need a bath after they eat." And they settled down. Barbara put a small cup of juice into Hodari's hand, and with a little help he held it and lifted it to his mouth, the juice dribbling down over his chin and onto a stomach that rounded out over his diaper. He concentrated intently on drinking. Could Patty do that? I wondered.

The women put food into their own mouths first. "They're fascinated by other people chewing," Sherry said as Mopey stuck a finger into her mouth. "I think it stimulates them to eat more." The gorillas ate, sucking the baby food off the spoons; the women wiped their chins after every mouthful. Hodari, quiet as ever, slurped his food around in his mouth. Mopey, on the other hand, gulped his down and immediately demanded his next bite, reaching for the spoon.

They both looked huge, with full coats of shining black hair, their full stomachs protruding over their diapers. And in my mind Patty seemed to shrink by comparison. She became grayer and feebler. I pictured her clinging to Lulu to nurse or begging to be picked up. She was different from either of these two boys in every way. Next to them she was delicate in her behavior and appearance. "Gentle" was the word that described her.

My mind came back to the present, Mopey's thrashing caught my eye as he reached for his cup. He insisted on holding it himself.

He drank thirstily and twisted around, his eyes roaming the room sharply. He was, I thought, planning his next move. But there wasn't time. Before he could squirm out of Sherry's grasp, she stood up. "Now," she said emphatically, "you've got to be changed." She turned to me again. "It's the third time this morning," she said wryly.

I followed them into the bathroom. Sherry laid Mopey on a table there; remarkably he lay still as she took off his dirty diapers. "You see how tiny his penis is," she said. "That's one of the reasons it's so hard to sex them. . . . Oh, Mopey!" He had reached toward the baby powder the moment she had turned toward me, knocked it over, and spilled it onto the floor.

After considerable effort he was finally diapered. Sherry lifted him, carried him to the hall, and deposited him in the playpen. He stood up against the bars and fought the mesh cover that Sherry lowered over him. "It's the only way to keep him inside," she explained. He looked up at her through the mesh and pushed his fingertips through it.

"Now you settle down and go to sleep," she ordered. But Mopey just stood there, straining to watch as Sherry diapered Hodari—with far less effort and excitement, I noted. Then she carried the tired little gorilla, freshly diapered and nodding on her shoulder, to her bedroom, where his crib was kept. The last I saw of Hodari he was still staring solemnly at me over Sherry's shoulder: the same wide-eyed, elusive, almost blank expression he had worn nearly all morning.

Yes, the Bronx Zoo had everything. But it could be a long time before Patty would visit the boys. For now she was alone.

CHAPTER 27
Patty Alone
Seventh and Eighth Months at the Bronx Zoo

The thread had been broken. In the next nine weeks I was to see Patty only ten times, and then for only a few hours at a time.

The day following the decision to keep Patty at the Bronx Zoo, I telephoned Brad in order to get permission to see her. He referred me to Dr. Dolensek. "He's in complete charge," he explained. Dr. Dolensek said only that Patty was already gaining weight, that she was sleeping a great deal and was still in a state of shock.

I spent my days with Lulu and Kongo wondering how Patty was reacting to her new circumstances and what her life now was like. The rumors flew back and forth.

"She's staying with the two other gorillas at Mrs. King's house," someone whispered to me confidentially. But I knew that Sherry King had left for Africa the day before Patty's accident.

"The Bronx Zoo says she was almost dead when they got her!" someone exclaimed in horror.

One of the keepers from Central Park told me that he had seen her. His confiding voice was full of awe. "And there's everything, even an oxygen tent if she needs it in a hurry. They treat her just like a real baby with toys and everything." He lowered his voice and looked at me in a furtive, confiding way. "She's much better off up there," he whispered. He looked around furtively to see if anyone else might have heard, then hurried away.

"I hear the Bronx Zoo wants to keep her," Andrée said to me. She looked at me wistfully. "Ah, Susie, do you think they can? It breaks my heart to see Lulu so alone."

The keepers went about their work as usual, but with leaden sadness. "Did you hear?" someone would say. "They say that we didn't take care of her." I shook my head in sympathy with them, for I knew just how much each of them had loved her. Zoo personnel try not to get emotionally involved with the animals they care for. It can be too heartbreaking for them when one dies or is injured or sold. But none of them could help loving Patty.

Each day I called Dr. Dolensek and asked what progress Patty was making. The cast had been removed, I was told, and a sling substituted for it. She was still sleeping a great deal, far more than was normal, but she had already put on so and so many ounces. Finally, on March 26, five days after the accident, Dr. Dolensek gave me permission to see her.

"Come in around one thirty," he said.

She lay in a playpen at the back of the room. How tiny she looked, a feeble figure in diapers, black on white sheets. She was staring up with wide and solemn eyes at a woman who bent over her and helped her up into a sitting position. Then I could see the sling that now held Patty's right arm. It looked like an Ace bandage that was wound around her arm and was then attached to a collar around her neck. She could move her arm from the shoulder, but not from the elbow, where it would have caused damage. At the moment she was not moving her arm, however. She was not moving at all. She simply sat unsteadily and expressionlessly in the middle of the white-sheeted playpen. There were toys strewn around her, splashes of unusual color in the sterile setting. She paid no attention to them, but kept her round eyes riveted to the woman's face. The woman reached down, patted the top of Patty's head in a reassuring gesture, said something to her, then turned away.

She saw me and smiled. "You're Susan Green," she mouthed. I nodded. She pointed to a counter by the window and indicated that she was about to make some instant oatmeal for Patty. "I'll be out in a minute," she gestured to me.

190

In the meantime, Patty had toppled down into a prone position. Her eyes had still not left the woman. The woman turned, said something to her, then opened the door that led from the room. Patty twisted around to look at the closing door, and I could see her little lips pursed in a whimper. Her eyes grew even more anxious.

"Hi, I'm Caroline Atkinson." The woman smiled as we said hello. "Please call me Caroline." She had a very gentle smile. We turned to the window to look at Patty.

"She's much better than she was when they brought her in," she said. "She was miserable, poor thing." *She looks miserable now*, I thought as I gazed through the glass at the lonely little figure. She was not the Patty Cake I knew.

"She's already gained a whole pound." Caroline's voice was filled with pride. She glanced toward the hot plate. The oatmeal was cheerfully bubbling away. "Excuse me," she said, "it's lunchtime."

The moment she reappeared inside the room Patty seemed to calm down. Once more she lay still, watching Caroline as she put the oatmeal in a bowl and then took out a spoon and a little pink bib.

Caroline settled down on a stool just in front of the counter with Patty on her lap. She dipped a metal spoon into the oatmeal and then tasted it herself. As Patty stirred restlessly in her arms, I could see the words, "No, it's too hot for a little gorilla." And then: "Open your mouth, that's a good girl." Patty opened her mouth like a little bird, and the spoon, brimming with oatmeal, slipped in. Patty pushed the cereal around her mouth with her tongue, dribbling a great deal of it onto Caroline's white laboratory coat. She really didn't seem to be interested in the food at all and finally swallowed a fraction of it. "Patty doesn't like it," Caroline had told me. "But she won't get any milk until she's eaten." Again a spoonful was lifted to her lips, and again dutifully Patty opened her mouth. She never once took her eyes from Caroline's face. Her expression never changed.

After each mouthful Caroline wiped Patty's chin with a little white towel. Not a drop of cereal was left on Patty's frowning little face. It seemed very strange and unnatural to me. I was so used to seeing her muzzle covered with a mustache of farina or oatmeal.

191

Seeing her wiped clean was somehow saddening to me. It brought home the reality that she had left the natural world of the gorillas and had truly entered the world of humans, required to become a part of it.

She had obviously accepted the spoon. The few times the keepers at Central Park had attempted to feed her, she had clenched her teeth and struggled to escape. It was only when they held the food in their opened palms that she had come forward to investigate.

Now her eyes steadily watched Caroline's face as the spoon made trip after trip and the cereal dribbled over the once-clean pink bib. Caroline talked to Patty all the while. And when the last spoonful had been swallowed, she reached for a bottle of milk and Enfamil. For the first time Patty showed a sign of interest. She reached out. Caroline held the bottle for her; Patty's one free hand rested on the bottle, too. As she sucked, her fingers bent and unbent in the same clutching movement she had used at her mother's breast.

She drank solemnly, her brown eyes focused on the blue ones that gazed calmly down on her. When Patty was finished, Caroline burped her, wiped her clean, and took off the bib. When she tried to take the bottle away, Patty clasped it fiercely to her chest and tried to turn away. But Caroline pried it out of her hand. The meal was over.

They were sitting directly in front of me at the window. Caroline held Patty up for me to see. She pointed out the place on her arm where it had been fractured and indicated to me that it was healing nicely. Then she pointed at me. "Look who's there!" she seemed to be saying to Patty. "Who's that?" But Patty would not look my way; she only struggled and twisted in order to keep her eyes on Caroline.

She was *not* the same Patty Cake I had known. There was something disturbing in her facial expression, something worried. There were new furrows between her brows, and her mouth was always drawn back just a little, as if in a perpetual soft cry. Her eyes had lost that wonderful eager expression they had always held, the curiosity and the interest I had always assumed to be part of her personality.

As Caroline held her, Patty began to fret and wiggle, perhaps a

192

bit uncomfortable in midair. Once more Caroline shifted the despondent infant toward me. Suddenly Patty's eyes were caught by the glitter of the gold heart I wore around my neck. She stared at it, looked away for an instant, then stared again. Her worried eyes traveled up to my face. It was only a glance at first, but within moments the glance was transformed into fascination. Her worried expression dissolved, and her eyes grew round. She gazed into my eyes as if caught by them. Suddenly she looked more like herself. She was familiar to me. She recognized me.

Caroline smiled at me through the glass. "She knows you."

Patty reached out to me. And I leaned forward.

"Hi, Patty." The words came out softly, although I knew she couldn't hear me.

Patty stared and stared. It was a long time before her head nodded sleepily and her eyelids drooped. Her body relaxed into sleep. Suddenly she stirred herself and tried to open her eyes; her lids fluttered for a second or so, fastened on me, then closed again. She could not stay awake.

Caroline laid Patty in the playpen, then bent over to change her. Patty half woke from her sleep, watching drowsily as Caroline diapered her. Then she once more drifted off to sleep, clutching an edge of a pastel blanket with one hand. Her mouth moved in a little sucking movement as she slept. She looked small and still, strange in her new world.

I had a little time now to look around her new home. It seemed very sterile: pale-green walls, red tiles, fluorescent lights and stainless steel counters. There was a scale and I now realized that what the keeper had described to me as an ozygen tent was an incubator. This was the nursery where animals that had been abandoned by their mothers were first kept. There was a table and a chair for Caroline and, of course, her cot, now folded and piled high with linens and her clothing. There were a few toys neatly stacked on the stainless steel shelves, Pampers, food, and formula. I thought of the cages in Central Park, warm with the bodies of Lulu and Kongo.

I was awakened from my reverie by the reflexion of someone in the window. I turned to find a bearded young man standing next to me. Behind his glasses an amused little smile flickered in his eyes.

It was very disconcerting. I wished I knew what the joke was. But he gave no hint.

I turned back to the window.

"Well?" he said.

"She looks fine," I said. There was a long pause. Self-consciously I took a few more notes. He didn't move but remained behind me, watching.

"She's so quiet," I said finally. "And she's sleeping so much. Do you suppose that she's still in shock?"

He put his hands in his pockets. "She's sleeping most of the day. But she's a lot better than when she came in. You know, she's already gained a pound."

I nodded.

He went on. "She was full of parasites, three different kinds."

"Can you tell me what they were?"

"I can get you a list, but it really isn't important." It was beginning to dawn on me. This was Dr. Dolensek! "She was obviously mistreated by the mother. Her musculature is underdeveloped for an animal her age, and she was terribly undernourished when they brought her in. Of course, she'd have to be in that condition. The mother obviously considered her a parasite."

"*What?*" My voice must have been shrill with surprise and horror.

"Well, she wouldn't allow the infant to eat anything."

"Where did you hear that?"

"One of the keepers from Central Park was here. He told us how Lulu always took the food away from her and refused to let the infant nurse. Of course, she was a dreadful mother. The baby has a calcium-protein deficiency. . . . "

As he talked on, I looked at him with growing amazement. "It's not true," I finally blurted out. "Who told you such a thing? Patty ate a lot—far more than Lulu did. You can look over my notes, it's all—"

The telephone rang, and Dr. Dolensek was called away. He hurried down the hall to his office. I turned back to Patty Cake, terribly disturbed.

I watched Patty sleep. Then, too quickly, it was four o'clock and time for me to go. It seemed to me that I had seen little and accom-

194

plished less. And what I had seen and heard had been disturbing. Slowly I packed my gear. Patty began to stir. Her eyes blinked sleepily a few times; her arm moved against the white sheet and pushed the blanket aside. Her eyes opened, and she glanced quickly toward me. She saw me there and fastened her eyes on me, even as Caroline leaned over her.

I stayed as long as I could, knowing that her eyes were on me and not wanting to leave. She seemed so solitary in this strange and cold place, removed from everything she knew. She was acted upon, not active, and helpless. She lay motionlessly in the green-barred playpen and showed no interest in any of the things around her. Only her eyes seemed really alive, and they were solemn now.

It was time to go. Her eyes, worried and wide, still focused on me. They followed me as far as they could before the solid wall came between us.

CHAPTER 28
Subsequent Visits
Eighth Month at the Bronx Zoo

A week went by before I saw Patty again. When I arrived, she was taking her afternoon nap. She lay on her back, snug in her playpen, dressed in a white undershirt and Pampers. Her good arm stretched out over her head. Caroline sat by her playpen reading. The moment that she saw me she waved, smiled, and got up to tiptoe out of the room and join me at the window.

"Hi," she said, "how are you? I was wondering when you were coming back." Her face was open and friendly. "Patty's taking her nap now. She had a huge lunch, and she's exhausted."

"How is she?"

"She's up to nine and three-quarter pounds now. Isn't that wonderful? She's gotten so big that her diapers don't fit any more. When she wets, it just slips out the sides and down her legs."

"That's terrific."

Caroline's face grew soft. "She's the sweetest little gorilla." Her voice was ecstatic. She had obviously fallen for Patty Cake, just as everyone else had.

"She has her mother's temperament," I explained. "Lulu's the same way."

"I just couldn't believe the way she adjusted to wearing clothes," Caroline said. "You should see her in her little pink stretch pajamas at night."

Our small talk was interrupted as Patty stirred, rolling over onto her belly. Caroline went in to her. She bent over Patty, gently trying to nudge the little gorilla awake. But Patty woke slowly, stirring a bit, moving a bit, finally turning over and opening her mouth in a yawn.

I was glad that Caroline was taking care of Patty. She had so much obvious delight in the little ape and felt so affectionate toward her. She was probably one of the gentlest people I had ever met. Her sweetness was apparent in every movement, even in the way that she walked. And her patience was unbounded.

Patty tried to put her right hand in her mouth, but the sling prevented her. She settled for a foot. Idly she reached for a rubber toy, let it fall, and began to wave her arms and legs in the air above her. Quietly she played with herself, watching her arms and legs floating above. She put her hand out to the side and touched the plexiglass that lined the playpen. She seemed to be playing with her own reflexion for a moment. Then she turned onto her side. She sat up, and then lay down again, quiet and content simply to lie there. She reached over and dragged a little transistor radio to her, hugging it.

Caroline reached down and picked her up. She carried her over to the window, and now I could see that the little worried expression was still there. Patty clasped the little radio to her chest. Caroline gently took it out of her grasp and laid it down on the table. "Look who's there," she said to Patty, pointing at me. "Who's that?" Patty was more interested in trying to climb up onto Caroline's shoulder. But Caroline refused to allow her to climb and shifted her around so that she was once more nestled in her arms. "Who's that?" Patty wanted to trace the edge of a button on Caroline's coat with her fingers. Again Caroline shifted her around. Patty reached up and put her finger against Caroline's face, exploring it. She was very different from the withdrawn gorilla I had seen a week earlier. Interest had returned to her eyes, and energy to her body. It was heartening. Patty was recovering not only from her injury, but also from the trauma she had suffered. Not that the distressing frown had disappeared; I suspected that would take a long time.

"Who's that? Look who's there!" And she saw me.

198

Again there was a visibly growing awareness as she stared at me through the window; recognition changed the expression on her face; her whole attitude seemed different almost immediately. She reached one arm out toward me, her fingers spread wide. Even her eyes seemed to grow brighter.

"Oh, Patty," I breathed.

When Caroline again carried her to the playpen and deposited her there, Patty's glance flickered toward me. And while her surrogate mother picked up her toys that had been strewn over the floor and washed them, Patty sat watching me. She clutched the little transistor, and although I could not hear anything, it seemed that she was listening to it. She sat very still, sneaking a look in my direction now and then. But as Caroline washed the floor, she caught Patty's attention, and Patty became intent on watching the familiar activity.

As the afternoon continued, it became clear that Patty wanted to be more active than she was permitted to be. She struggled to stand up in her playpen and put her left hand over its rim, trying to hang one-handed, and then shifted around, trying to climb up. When Caroline fed her, she tried again to stand up on her lap or to climb over her. When her meal was finished, she wanted very much to explore and struggled to be free. But she was relegated to the playpen while Caroline went about household duties. Again Patty clutched the little radio to her chest.

The radio had become important to her. The only time that Caroline ever left Patty was to do the laundry. But Patty needed the constant reassurance of life around her. There seemed to be a terror that enveloped her when she was alone or when, despite someone's physical presence, she felt alone. At these times she became hysterical. Then Caroline discovered that the noise of a radio would calm her. Patty was now as dependent on that radio as she was on Caroline. She clasped it tightly to her chest or dragged it along after her as she crawled. She listened to it as she fell asleep, and if it was removed, she screamed for it. The radio was now the first thing that she reached for when she woke. It did not matter what she listened to: symphonic music, rock 'n' roll, or the human voice. It was sound, and it comforted her. Before she left the room, Caroline turned on the radio and let it play softly.

Caroline's extreme gentleness was so unlike Lulu's matter-of-fact handling of Patty Cake that I wondered if Patty was not sometimes frustrated by the restraints of gentle hands. When Patty showed an inclination and an ability to climb, Caroline did not encourage it; she did not tumble the baby or swing her as Lulu might have done. As Patty stood and hooked her left hand over the railing of her playpen, Caroline unclasped the hand and sat her down. Patty was not exercised. Rather, she was held.

Her arm was healing rapidly. Caroline proudly showed me the callus that was forming over the site of the break. I could not help wondering whether if there had really been such a bad calcium deficiency when Patty had been injured, the arm would be healing as beautifully or as quickly as it was. Dr. Dolensek was more than pleased with Patty's progress, and he thought that the sling would probably be removed in three weeks. He spoke to me briefly about putting a little set of stairs into the nursery for Patty to exercise on.

I seldom saw Dr. Dolensek. Sometimes he appeared reflected in the window glass as he had the first day I came. He made a few comments about Patty's weight or general condition, then was either called away or hurried to an appointment with a sick or hurt animal. I sometimes thought that he was curious about me, but he never asked what I was doing or why.

"Lie down and go to sleep," Caroline instructed. But Patty kicked off her light blanket, not wanting to go to sleep. She waved her arms and legs, even the broken one, as fast as she could. She began to chew on the little rubber toy, and she scratched her leg, turning over and staring into space, playing with her hands, anything to keep herself awake. But slowly her eyes closed, opened, then closed again. As she slept, her arms and legs moved and twitched as if she were dreaming.

It was again time for me to leave. The raindrops spattered heavily on the skylight, sounding like the rapping of drums overhead. Perhaps, I hoped, the downpour would delay my going . . . at least until Patty woke. While I packed my gear slowly, one of Dr. Dolensek's assistants wandered over to me. "It's too bad you have to go out in this," he said.

200

"I sure didn't expect it," I said, hinting just a bit. "I didn't bring an umbrella."

He looked sorry for me but only shrugged. And I continued packing. And as Patty still slept, I waved good-bye to Caroline and started down the hall. Dr. Dolensek was in his office. I always felt so awkward and silly when I spoke with him, but I stopped to ask him when I could return. He replied that I should call and make the arrangements, but he thought that it would probably be all right the following week.

But when I called, I was told that it was inconvenient that day. The following day I called again. And the next.

The third time I was able to see Patty, her arm was out of the sling. It had happened ahead of schedule, and everyone was very pleased. Now I could see how she used her arm. Perhaps it would mean she could return to Central Park sooner than anyone thought. I was very excited and eagerly began to unpack my things. I had just taken out my pencils when a phone behind me rang. I barely paid any attention when the assistant answered it. Patty had just seen me and was reaching toward me.

The young man approached me and stood by me as if waiting to say something. He seemed a bit embarrassed.

"Uh . . . that was Mr. Conway, the director of the zoo," he said. "He wants you to leave."

My heart sank. And I knew that the disappointment and the hurt showed in my eyes.

"I'm sorry," he said apologetically. He tried to explain. "Mr. Conway didn't even know you were here." He shuffled his feet. "It's something about the newspeople. Someone called and said that if you were here, why couldn't he be?"

"I see." My voice must have sounded very disheartened.

"Sorry," he said again, then turned around and went into the laboratory across the hall to get back to his work.

So, with one last look at Patty, I again packed up my pencils and papers. And left.

201

CHAPTER 29
Transitions
"We Miss Thee at Home"
Seventh Month at the Central Park Zoo

I remember seeing a gravestone in a very old graveyard in Boulder, Colorado. It stood thin and weatherworn over the grave of a little girl who had died during an influenza epidemic of the mid-1800's. The inscription read simply: "We Miss Thee at Home." I had forgotten all about it until now. Somehow Lulu's solitary and idle figure reminded me of it.

Lulu's grief did not end on March 22 or 23. It lingered far into April, diminishing slowly.

Lulu was despondent and subdued those first days. She often sat in a stupor, staring into space, motionless. Sudden frenzied movements interrupted her lethargy, and she swept the cage as if she were searching for something, only to drop again listlessly to the floor.

Too, she showed unaccountable irritation which would suddenly dissolve in a sigh. And all the time Kongo watched over her and seemed to tend to her, sensitive to her moodiness and her grief. Even a week after Patty's departure, as the two gorillas sat on the crossbar, Kongo reached out his foot and barely touched Lulu's thigh. She whirled around and pounded him, catapulted to the bars above, swung down toward him, whirled away again, then abruptly sank down next to him. She leaned on him heavily, her head resting on his chest. He looked down at her and did not move.

It was that afternoon that Kongo began to make sexual advances. Lulu seemed to tolerate them for a little while. But she soon became irritable and ill-tempered. As Kongo again came toward her, she drew her mouth back just enough to show her teeth. Her mouth moved as if she were muttering, but without a sound. When he took another step toward her, her eyes narrowed. He shifted his focus to a nearby apple.

Kongo was becoming frustrated. Soon he stalked into the next cage, and I could hear him banging the walls there. The metallic boom reverberated through the house, rising in intensity, gaining in momentum. Lulu heard the commotion and sat still for a few moments. Then she rose, hooting. The sound became a scream; she bounded toward Kongo and disappeared into the cage. They rushed out. She chased him, screeching, hitting out at him when she could, pounding him when she backed him up against a wall. Just as suddenly her fury ceased, and she dropped down onto all fours. Kongo came toward her slowly. He simply put out his hand. Again she turned on him. A few minutes later he tried again. She began to hoot irritably; he seemed to have had enough. He put his arms around her, heaved her off the ground, then threw her to the floor. She landed heavily; he backed away. She picked herself up off the floor, and he watched her from the corner of his eye. She limped wearily toward him, then stopped and slowly turned. She walked away.

She was covered with the sores and scratches she had received on the day of Patty's accident. They were on her face, her brows, her foot, her leg, and her lip. As she moved away from Kongo, walking lamely and stiffly, she seemed beaten, the sores bright and liquid against her dark-gray skin and black hair. Kongo watched her closely; then, as she again turned toward him, he deliberately strode into the next cage and left her alone. She seemed rejected. She wandered aimlessly, sat motionlessly. Absentmindedly she played with an apple which lay at her feet, but made no move to eat it. It finally dropped to the floor and rolled away. Since Patty had been taken away, I had not seen Lulu eat anything. She had merely drunk sparingly from the hose a few times and had accepted a little milk.

Kongo peeked in the doorway, glanced at her, stalked in, and went to the very center of the cage, where he made an elaborate show of dusting off a place for himself. He then reclined. Lulu glanced toward him; then she went and sat down by him. When he did not move away, she began to groom him, halfheartedly parting some hair on his arm. It looked, I thought, as if she were making a thin attempt to be social. For a long while her fingers slowly separated the hair on his arm or his back, and she bent in close to peer into the part she had made. He allowed her to groom him, but otherwise ignored her. Finally, her small, quiet movements stopped, and she rose in a tired, hopeless way and walked away. She fiddled vaguely with some sweet potatoes, then timidly approached him again. Again he disregarded her attempts to be friendly.

She lay still for nearly the entire afternoon in a familiar pose, her arms folded across the chest, her legs drawn up in the cradle position. Her head rolled back and forth, and her left hand twitched; she stared purposelessly. Once she reached down as if to move the baby from her lower belly to another position. But in midair, her arm stopped and hung for an instant before she put it across her chest once more.

She did not look at me although she lay at the front of the cage a few feet from me. Once, as her glance fell on me inadvertently, she turned her head away and actually covered her eyes with her hand. Then she rolled over and turned her back to me.

The following day I could see a slight change in Lulu's attitude. A little interest was aroused in her, and she began to come out of herself. I could not help thinking that Kongo's change in attitude, his lack of tolerance for her moodiness and unpredictability, was in some part responsible for Lulu's return. She needed his attention, it seemed, and once he began to deprive her of it, she began to come around. As soon as she did, Kongo was more than willing to play. They swung a bit, together, for the first time in a week. They both received showers. She went to the trough to splash the water and to gather old bread and vegetables as they floated by. She sat

205

squeezing the water from a handful of soggy lettuce. When Kongo came by and tweaked her stomach hair, she did not fly into a rage. A few minutes later she reached for some cornmeal and ate. I felt a surge of energy run through me. The turning point had come. Now it was simply a matter of time. The initial shock seemed to have lessened. Slowly, as the days passed, she reacted more and more to things outside herself.

It was in April that her restlessness began to change from the frantic distress she had first exhibited to one which rose from boredom. For her role had been taken from her when her daughter was taken from her. Without the role of mother, time stretched out relentlessly and uselessly; she found no adequate substitutes for her energy.

There were many signs of nervousness and restlessness. She continually looked into whichever cage was adjacent to the one she occupied, as if still searching. Often she jumped at any loud or unusual sound, and her head swiveled, her eyes wide and intent for an instant. She moved continuously from place to place, only to settle down, then to spring up and begin to roam again. She hugged the walls as if avoiding the center of the cage. I wondered if she were still searching for Patty along the bars and the walls or if they gave her a sense of security that the open space of cage could not.

Their lives became more normal. Eventually only a few reminders of Lulu's reactions to her loss remained. These episodes came without warning, and I could find no clue to what had set them off. In the midst of playful activity, during a feeding, chase, or rest, Lulu would suddenly suspend all action. Her head tilted to one side. She stared unblinkingly and, it seemed, unseeingly. Her lower lip hung loosely. Her body became motionless. She would sit that way for as long as five or seven minutes, a strange, suspended posture.

Suddenly she would shake herself out of it; a violent movement shook her head, her eyes seemed to focus on something real before her, and the vacant look disappeared. Her body seemed able to function again.

Sometimes this transition occurred slowly. As she came out of her trance, for that was what it seemed to be, her head moved almost imperceptibly until it was held upright and she began to act more normally. Her face often twisted into a grimace as she rose. And at the same time another action developed: a strange wild dance. She bounced up and down, her arms flopping strangely around her, her head bobbing and shaking violently, her lower lip swinging loosely.* The dance seldom lasted more than half a minute, although it sometimes continued in fits and starts. Then it, too, dwindled in its convulsive intensity, and in a few moments Lulu would go off by herself to sit, kneading her shrinking breasts, or to play with something she had found. Often she played with her fingers, staring at them as she twisted them, dovetailed them, and entwined them over and over.

Lulu's behavior often seemed to me to be like a flash flood which intruded into her usual calm and then abated, a sadness that spilled into the atmosphere of the cage, then slipped away.

As Lulu became less moody and more active, Kongo became more assertive. As Lulu's importance, her motherhood-dominance status, diminished, Kongo's dominant role grew.

Since the baby's birth, Lulu had been dominant in many ways. She had often taken the choice position in the cage and had not been threatened by Kongo as he came toward her. She usually ate what she wanted and did what she wanted with little or no interference from the male. She had ruled any situation that involved the baby. She could snap at him, even hit him. And his only reaction would be to run. He tolerated her moods, gave way to her desires. It was in the natural order of things.

But now the baby was gone, and Lulu's grief was subsiding. Kongo's role as leader and protector of this little troop of two remained, while her role and her purpose were gone. Perhaps Lulu's admission of submissiveness was the turning point in their relationship.

Kongo seemed to take possession of objects, claiming them as his own. Lulu obediently let them go. Lulu had not suddenly become a shrinking violet; she still scrapped with him and hooted and

---

*This was an activity which Lulu had performed often since the birth of the baby, an activity which I called the Watusi after a popular social dance of the time, remarkably similar in its movements. It could also be performed sitting, eliminating the action of the legs, and lying down, in which case the primary movement was in the head and the shoulders or sometimes in the arms.

207

screeched. But there was no baby for her to sweep away from under his nose, no particular thing that was undisputedly hers, no responsibility.

Kongo claimed the chain that swung from the door of the cage and the tire and the swing and the new bench which Fitz had installed in one of the outdoor cages in anticipation of Patty's return. Kongo could claim both the bench and the swing by sitting casually on the platform which ran down the center of the cage and holding onto the swing above or by sitting on the swing, enjoying its swaying motion and the view, ready to bound to the platform should Lulu show signs of resting there. But she never did. She climbed over its back on her way to and from the adjoining area, hustling over it quickly, keeping an eye on Kongo above. Only when he was nowhere in sight did she allow herself the luxury of sitting comfortably at the front, leaning her legs on the bars. When Kongo stuck his head in the door, she would glance at him, then move. Sometimes they seemed to be playing a form of musical chairs. Kongo followed her from place to place, displacing her from one choice resting spot to another.

Even when Lulu discovered something first, Kongo usually took over as soon as he realized there was something entertaining or valuable in her possession. At the beginning of May Lulu discovered that by scraping and digging she could loosen one of the bricks in the wall of her outdoor cage. She worked with concentration and perseverance. She was absorbed by the challenge of getting that brick free from the wall.

As soon as she held the brick in her hand, Kongo appeared. In a moment he had sent her to the doorway, leaving the brick behind on the floor. She watched as Kongo picked up the brick, then dropped in on the cement floor. He held it and scratched the floor with the edge of the brick, then put his foot on it, dropping his toes over the edge. He looked at the floor where he had scratched it, and he must have seen something there, for the next moment he picked up the brick and scratched the wall. He looked very closely at the red marks that he made.

As Lulu watched, inching forward bit by bit, he put the brick on the floor and pounded it, turned it over and pounded the other side. He took it back to the spot where Lulu had pried it loose and examined the space. He sniffed it; then he ran his fingers around the ragged edge of the remaining cement and neighboring brick.

Lulu crept up behind him, watching intently. As he saw her shadow or felt her presence, he suddenly threw up his arm in a threatening gesture, and she scurried back to the doorway. It was Kongo's brick.

When Richie realized that Kongo had it, the game became more fun. For Richie would try to retrieve it. He tried to bribe Kongo with a raspberry ice, but the ape merely switched the brick from his hand to his foot, sliding it on the floor away from the keeper as he reached for his ice. He actually picked up the brick and tapped it on the bars, inches in front of Richie's face. But when Richie made his move, Kongo moved. He strolled off, the brick in one hand, the raspberry ice in the other.

Now Richie got smart. He offered a lemon ice to Lulu, hoping that Kongo would drop the brick and leave it behind him in his eagerness to take the ice from his mate. But Kongo merely put the brick behind him as he ran for the ice. He ended up with both, while Lulu stood empty-handed and powerless.

Kongo teased Richie with that brick, tapping it against the bars, holding it out toward him. Each time Richie grabbed for it Kongo was faster. Each time but once! Then the game was over.

Both Kongo and Lulu stood at the edge of the bars, looking down at their brick. It lay, discarded and temptingly close, on the cobblestones at the base of their cage. Lulu was the first to leave. She headed toward the door, and there she looked back at Kongo. He glanced up toward her. I could almost swear that there was a flirtatious wiggle of her body; she tossed her head, then turned and ambled inside. Kongo stood staring at her slight feminine form for a few seconds, then followed her in.

CHAPTER 30
Lulu in Love
April 11, 1973; Eighth Month at the Central Park Zoo

Spring was full-blown and full-blossomed. The cherry trees were sprouting lovely pink petals. The skies were blue and warm. On April 11 Lulu went into her monthly cycle.

Gorillas are not monogamous. When the female goes into estrus, she approaches the male of the moment and presents herself to him.

Lulu was in love with Richie.

When Richie appeared before her cage, she was off, following him up and down with lost, wide eyes as he walked in front of her. If he disappeared into the keeper's room, she waited for him, and then, when he came near, she reached out toward him, begging to be petted, talked to, stroked. She sat staring at him with lovesick eyes, while Kongo watched her from somewhere in the background, glowering.

As Richie stopped to pet her arm and crouched down to talk with her, she stood up and bent over. This was Kongo's chance. He came up behind her stealthily, grabbed her hips, and drew her back against him. He began to thrust, panting a little. But Lulu hardly seemed to know he was there. Her eyes were still fixed on Richie, kneeling in front of her cage. At this point Richie decided it was time for him to leave the two gorillas together. As he walked down the aisle of the keeper's walk, she followed him. For a moment Kongo hung on, but then he gave up and let her go. She jumped up onto the crossbar, and as Richie appeared beside me, her eyes narrowed. Was it possible that she was jealous?

As soon as Richie left me, she turned to Kongo. They played, pushing, pulling and hugging each other, their mouths open in the physical enjoyment of each other. A lovely wrestling match was beginning to grow. Perhaps it would lead to other things.

But then Eddie came over, and once more Lulu abandoned Kongo for a man.

Kongo soon intervened and physically took Lulu to the back of the cage. But as soon as she could get away, she went trotting back to Eddie.

"What do you want, my sweetheart?" he cooed, teasing her. "You're a good girl." She put out her hand toward him and tucked one of her fingers under his sleeve. "You turn around and I'll scratch your back." Kongo, glowering at them from a distance, finally stood up. He stalked toward them, his head down like a ram's. Lulu backed away. Kongo charged. Eddie left.

Kongo kept his eyes on Lulu and displayed his strength. He smacked the tire that hung from the ceiling bars so that it swung up, hit the bars, and plummeted down till it jerked at the end of its chain. He charged back and forth across the floor, excited, stopping to rise and beat his chest with rapid booming sounds that reverberated throughout the house. He stood at attention, his back arched, his head high, his snoot thrust forward. Lulu stood still as he came up behind her and mounted her, one leg lifted heavily over her rump. He lifted his weight onto her back, wrapped his legs around hers, his arms around her neck. His tremendous bulk and weight lay heavily on her small form.

Then she spotted Richie.

She staggered forward toward him, swaying and faltering under Kongo's weight. Even as she moved, Kongo began to thrust. His body weighed her down, and her slim back became almost concave, but she laboriously tottered toward Richie. Finally Kongo began "huffing" quickly; then he let out a long and powerful roar. When he dismounted, her body seemed to rise, almost floating into space. She shook him off impatiently, then bounded after Richie.

But he was already disappearing into the keeper's room. She sat down and waited for him to return.

Repeatedly Kongo displayed, running back and forth into the private cage, then out again toward Lulu. But if Richie appeared, Kongo did not have a chance. Lulu headed toward the keeper. She rose up onto her legs and beat her sacrum with her fists as she followed him. If Kongo came close, she retreated, and if he left her, she came again to the bars to pine after Richie.

Kongo and Lulu's sexual play continued all day; they mated continually. Although it was often Kongo who began the wrestling which would ultimately lead to copulation, sometimes she approached him and backed up to him, obviously inviting the sexual act, yet still watching for her human friend. While they mated, one or two of the keepers stood in front of the cage, egging them on and teasing, *Oooooooh, aaaah, oooooh.* They grinned and winked at me. *Ooooooh, aaaah.*

When it was over, Lulu stood erect and waddled toward them, tapping her sacrum as she came. Kongo sat at the back of the cage, watching the men, who continued to jeer.

Slowly Kongo stirred, heaved himself off the floor, and lumbered to the front of the cage.

Lulu retreated to the door. Kongo looked at the two grinning men. Then carefully, deliberately, and swiftly he swept back his hand. He sprayed the three of them with urine. As they swore laughingly, he turned and walked back slowly to his place.

213

CHAPTER 31
Ronnie
Eighth Month at the Central Park Zoo

For weeks we talked about Patty returning to the Central Park Zoo. We sat in Fitz's office and dreamed about it, stood before the cages and hoped. It was a fantasy. Yet perhaps it could come true. The more we spoke about it, the more we believed that it could—no, that it would—happen. Fitz began to prepare for the day when Patty would be reintroduced to her parents.

The first consideration would be to find someone to take Caroline's place. I secretly hoped it would be me, but that was a fantasy. No, the ideal person would be Veronica Nelson.

Ronnie was one of the few female keepers not restricted to working in a children's zoo or animal nursery. I had first met her in October, when she had come into the Lion House to see Patty Cake. She stood in front of the gorillas' cages, and Kongo strutted to the bars to stare back at her. Being relatively new there and feeling insecure so soon after Patty's birth, I envied Ronnie's hearty, matter-of-fact way with the animals.

"Hiya, fella, how're you doing?" Kongo put out his hand toward her. She grinned. "Oh, no, you don't!" Kongo sat back on his haunches and stared. She just stared back, grinning at him.

Suddenly a muffled chattering sound distracted me. A series of squeaks and squeals came from Ronnie's direction, and I peered at her curiously.

215

"What do you want?" She was obviously talking to something on her person. The squeals exploded again, then ceased.

"OK, OK!"

Ronnie reached inside the top of her blouse and drew out a wrapped diaper. The diaper quivered and jerked as another string of trills pealed forth. I looked closer. Nestled within the diaper was one of the tiniest monkeys I had ever seen.

Only its walnut-sized head emerged from the top of the folds, twitching this way and that, twittering and chattering. It wriggled its body until some of the cloth was pushed aside and I could see tiny toothpick fingers, attached to matchstick arms. They never stopped moving, but darted back and forth to clutch at Ronnie's blouse. All the while the shrill, birdlike sounds scattered into the air. Kongo stood watching, spellbound. So did I.

It was an infant female capuchin monkey. It had been rejected by its mother at the Brooklyn Zoo, where Ronnie worked at the time. Attacked by the other monkeys in the cage, it had been beaten and bitten until bloody and nearly dead. Ronnie managed to get it out of the cage and took it, limp and starving, to the children's zoo nursery. When the monkey was well, Ronnie took her home. She named her Darwin.

Dirty Gertie was the next addition to Ronnie's growing family. When she became a member of the household, she was two months old, just a month older than Darwin.

Gertie was an olive baboon. She had also been mistreated, but not by her mother. Passy (short for *pazza*, which means "crazy" in Italian) was a good mother. She had already brought up one youngster, a male named Andy, who had grown into a mischievous, rambunctious young baboon.

Baboons are not known for their gentleness. They are vicious, aggressive, and often bad-tempered, taking advantage of the weak or sick. Andy was playing true to form. His father, Big Boy, was worse. For some unknown reason he grabbed Gertrude from her mother one day and dragged the screaming infant across the floor. He flung her into the air, then caught her. He leaped to the top bars of the cage and dropped her ten feet or so to the floor. Passy bounded toward her and put the injured, terrified infant under her stomach. She tried to keep away from her mate, but Big Boy was

216

not satisfied. Once again he kidnapped the infant while the mother looked on helplessly.

Andy was not much help; he, too, joined in the fun. He dragged his little sister across the floor and pulled her up the bars by her tail. He hauled her across one of the pipes that served as a crossbar high in the cage while she screamed in pain. The mother could do nothing to prevent it.*

It took some time before Fitz and the keepers could rescue Gertrude from her tormentors. By the time she was removed from the cage many of us thought that she was dead. But somehow Gertrude survived.

As soon as her wounds were treated, they turned her over to Ronnie, who was now a regular keeper at Central Park. Under her compassionate ministrations, the hurts were mended; the fractured wrist and the torn face healed quickly. Only a few scars remained. Soon Dirtie Gertie was living an enjoyable and "normal" life with Ronnie and Darwin.

As she gained assurance, she began to behave like a normal baboon: impish, demanding, and aggressive. It was wonderful to see. Even her temper tantrums, her screams, and her threatening gestures were wonderful. That was the way she was supposed to be. She was becoming so baboonish that Fitz considered reuniting her with her family. The idea frightened me. I could not help remembering how pitiful she was when she had been taken from the cage.

But Fitz was a practical man. The day would come when Gertie was no longer cute. She would no longer arch her back and bare her teeth in playful aggression. Like her parents, she would become a dangerous animal. Ronnie could not keep her forever as she could Darwin. For Dirtie Gertie could never be a pet. It was not in her nature. Either she would spend the rest of her life in a cage alone, or she might live as a member of a baboon troop. It all hinged on the risks that Fitz was willing to take. Ronnie thought that Gertie could make it. Fitz believed in Ronnie's handling of the animal. He trusted her judgment.

In mid-March it was time to try. Over a period of days Gertie

---

*This is not extraordinary behavior for baboons. Jane Goodall has recorded and filmed similar occurrences in the wild.

was reintroduced to her family—first to Andy and then, when her fear turned to cautious aggression, to her parents. Within two days Gertie had begun to relax, and as soon as she was introduced to her mother, Passy took control and protected her. In no time Gertie was secure in the knowledge that her mother would come to her aid if necessary. Soon life was normal in the baboon cage; the reunion was a total success.

Fitz had been right. He gave the credit to Ronnie.

Ronnie now came to the gorillas' cages to become acquainted with Lulu and Kongo. If she were to care for Patty, give her bottles and vitamins, play with her, and examine her, Lulu and Kongo had to trust her. And Ronnie had to learn something about gorillas.

It was not as simple as being accepted by the baby monkeys. The adult gorillas were not dependent on her, and she could not, should not, be in control of their lives. Rather, she had to be accepted on their grounds.

She started coming to the Lion House at the beginning of April. Kongo would take one look at her and splash. Her hair had a red tinge to it, and Kongo always seemed to have something against redheads. No one knew why. Ronnie would stand in the doorway to the keeper's room and talk to him for hours. She offered to play with him, slapping his hands or pulling his fingers between the bars. But inevitably he ended their sessions by making threats, trying to grab her, or spraying her with urine. She ducked out of his way, then turned to me, asking why.

She tried everything. Often enough Kongo allowed her to stroke him or to grab a finger and play for a minute while he lazed by the bars. But as soon as Ronnie began to relax with their games, he took a swipe at her.

Then she began to think that the way to his heart was through his stomach. She not only brought him his usual foods, but thought up special treats he might enjoy. She bought strawberries. One dollar and seven cents a pint! She carefully laid one huge succulent strawberry on the horizontal bar. He strolled over, delicately plucked the berry from the bar, and popped it into his mouth. Then he sat down and opened his mouth, waiting to be served. He graciously accepted the berries, one by one. Lulu shifted behind him, hooting softly, her eyes glued to the fruit. As Ronnie moved over to give her one, Kongo charged. He chased Lulu away and then sat down sweetly, and with sleepy eyes, he opened his mouth for more.

When Lulu went into estrus, matters were further complicated. Despite Lulu's lovesick feelings toward Richie, she resented Ronnie's attentions toward her mate. As soon as Ronnie entered the house and Kongo started toward her, Lulu came running, barking out her objections. She backed Kongo into a corner, then took a stance between her mate and the woman. She stared at Ronnie suspiciously, then turned her gaze onto Kongo. Once she actually, unbelievably, charged Ronnie!

And all the time Ronnie kept up a running monologue with each of the gorillas, sometimes mocking, sometimes harsh, sometimes soothing, but always friendly.

As she offered Kongo a banana, his hand suddenly shot forward between the bars as if to grab her hand.

"Oh, no, you don't!" She shook the banana savagely at him. "You'd better be careful," she threatened, "or I'm gonna beat your ass!" Kongo looked at her blandly, then put both his fists against the bars. Obediently Ronnie began to play with him, slapping his broad hands. She scratched his back when he turned around and pulled the long hairs on his belly when he again turned toward her. Lulu glowered at them from a distance. "You are the handsomest little ol' gorilla," Ronnie growled at him. "You are the meanest . . . come here, give me your hand . . . you are the lousiest. . . ." she cooed at him, and he poked his nose against the grating, enjoying it all. Lulu darted toward them. She actually punched Kongo and shoved him against the bars. He looked at her furiously, chased her off to the rear of the cage, then turned back to Ronnie. In another moment she was once again offering him strawberries.

Soon Ronnie's designated time with them was up for the day, and she started out of the keeper's walk. Then Kongo pounded the wall and stared at her retreating form with angry eyes before he turned and strolled back to Lulu.

Two days later Kongo was absolutely obnoxious. Ronnie was trying hard to be accepted by him, for she felt that if Kongo liked her, Lulu would take his lead and everything would be very simple.

219

It was not working out that way. Kongo teased Ronnie, offering to play. He snarled at Lulu and sent her scurrying back out of the cage. Then, when Ronnie turned her back for an instant, Kongo swooped up some feces and hurled it at her. She ducked into the keeper's room just in time, and he missed. But she was shame-faced at having been tricked.

She tried to feed him; he threw urine at her. When she tried to play, he walked away. Lulu came up to Ronnie and for once stood still while Ronnie stroked her back. Kongo reached out and hit his mate. They began to vie for Ronnie's attention. Seeing trouble ahead, I knocked on the plexiglass window that separated Ronnie from the aisle where I stood. She turned, and I motioned to her to come out. "There's going to be a fight," I mouthed. But she did not understand. She turned back to the animals. Slowly the two goril-las blocked each other's access to Ronnie. They stood up, backed down, complained, and threatened, each of them maneuvering the other into position. Suddenly they were fighting! Screaming and biting, they rolled on the floor, grappling with each other. Kongo picked up Lulu bodily and tossed her to the ground. She landed heavily, then spun up to attack. She chased Kongo full speed across the cage. He hit the tire and set it spinning.

Somehow his aggression was averted from attack to display. He ran, pounding his chest until the sounds boomed, echoing each one, thundering, accelerating, bursting into the space. He charged and swiped a sheet of water against the bars. It splattered heavily on the glass before my eyes.

In the midst of it all Lulu stood still. Kongo came to a sudden halt and waited, poised. Lulu went to him. She turned and present-ed herself to him in a submissive posture. He lifted one leg over her rump for a moment and then strode forward to stand before her, face to face. His arm draped over her slight shoulders.

Ronnie had dashed out of the walk when the fight started. She came running up to join me and had watched the fight with dis-belief. Now I heard a long expiration of breath as she relaxed.

"Wow!" she exlaimed. "That was incredible!" She turned to me. "Did I do that?"

"Don't worry about it. You're trying too hard, that's all."

She looked upset.

"Really." I laughed. "The fight looked a lot worse than it was. Sometimes they play almost as roughly." I felt awkward. Ronnie really was wonderful with animals. It was simply that she did not yet know these gorillas well enough. But I didn't want to butt in. "They'll come around. It's just that they've got to accept you."

We stared into the cage for a minute.

"Look, Ronnie, can I say something?"

"Sure," she replied. "Anything'll be a help."

"Well, you can't make one of them jealous. And you really can't let Kongo take advantage of you. This business about "feed-me-and-I'll-play-with-you" isn't going to go anywhere. He's too smart. Why don't you just hang around for a while and let them get used to you? You've got plenty of time. Patty isn't coming home tomorrow."

From then on Ronnie relaxed. I was right; Patty wasn't coming home tomorrow.

CHAPTER 32
The Moat
April 30, 1973, at the Bronx Zoo

I was sitting in a cement arena that terraced down to a narrow canal. The arena was filled with barrels and boxes of toys, baby bottles, formula, and Pampers. At the bottom of the terraces a tall wall rose abruptly, on the other side were the blurred faces of an audience. Around me gorilla babies played.

Across the wide terrace of the moat Mopey charged a barrel. It toppled and rolled, the toys spinning out over the terrace. He chased after them as they bounced down the steps, his arms flung out wildly as he ran. Almost automatically Hodari gave a little squeal and reached out for Sherry. Best of all, coming toward me slowly, shyly, was Patty Cake. Her eyes were fastened on the sparkling white paper on my lap, and she put out her hand and gently touched it. Dark smudges scraped across the drawing.

I couldn't have cared less. Two sweet, warm eyes looked up into mine, and Patty Cake rested her hand casually on my knee. I leaned over to hug her; her arms came up, and she held onto me.

And it was real!

Two weeks before, just as I was about to step into the elevator, the telephone rang. "Oh, no!" I dropped my things in the hall and fumbled in the bottom of my bag for my keys. Why did this always

223

happen just when I was leaving the house? Now I would miss my bus! I flung open the door, listening for the phone to stop ringing. But for once it didn't, and I lunged for it.

"Hello!"

"Mrs. Green?"

"Yes."

"Just a moment please. Mr. Conway of the Bronx Zoo calling." I sat down. I couldn't believe it.

"Hello, Mrs. Green? This is Bill Conway. Commissioner Clurman's office asked me to get in touch with you concerning our mutual interest in Patty Cake."

It really was Mr. Conway. In seconds my impatience had turned to astonishment and then a heart-racing nervous tension.

"I understand," the voice was saying, "that you are interested in seeing her."

Interested?

For weeks I had been trying to get permission to see her. But the days had passed without any word, and I had given up. Now, out of the blue, Mr. Conway himself was asking if I was interested.

"Be calm," I told myself breathlessly. "Be professional."

"Of course," the sincere voice on the other end was saying, "I want you to understand that our refusal to you was by no means a personal matter. It is simply not our policy to allow people into the hospital except under very extraordinary circumstances for some very good reasons which you can well appreciate. As you are aware, perhaps, we have a large collection of exotic birds which may carry any number of rare diseases. It is important, you understand, for them to be quarantined until we are absolutely certain that they are healthy. And we must be very careful that no one is exposed to them. Our concern, you see, is really for your own protection."

"Yes, of course, I can see that, Mr. Conway."

"And of course, there is something else. You can see, too, that if we let one person in, we would be obliged to let others in as well. As happened in your case, by the way. I did not even know that you were there until one of the reporters called and said, in essence, 'If she's there, why can't I be?' And of course, he was right."

224

I could feel any pretense of calm slipping away. I might never see Patty again at this rate.

"And, too," the positive-sounding voice continued, "you must understand our position concerning Patty Cake herself. You know that she was a very sick little animal when she arrived here." He began to enumerate all that had been wrong with her. I was beginning to panic. When he paused for breath, I saw my chance and jumped.

"Mr. Conway, you know I would really like to talk to you about this. As you know, I did so much work with her in Central Park and I've hardly even seen her in the Bronx. You see, I feel that I am missing so much of her emotional adjustment. No one seems to be able to tell me anything about that, and really, it's so important! It's very frustrating not knowing what she's going through now. And I haven't been able to finish one of the drawings I started at the Bronx Zoo. I'm a very slow worker, and the only drawings I have of her there are so sad. It's not fair to the Bronx Zoo either. . . ."

I could hear myself rattling on and on. Why did I always babble hysterically when I was speaking with someone I wanted to impress? I stopped myself suddenly. Worse. There was total silence from the other end. The man must have been in shock, trying to figure out what I had said. I felt desperate.

"Mr. Conway," I heard myself saying, "I really would like to talk to you. Do you think we could make an appointment sometime?"

To my astonishment he sounded absolutely enthusiastic about the idea. "Yes, I think that would be delightful! Let me look at my calendar. I think that lunch would be best, don't you? That way Mr. House can join us, and we can really take our time. Would the thirtieth of April be satisfactory?"

As I hung up, I wondered what he looked like. His voice had been utterly charming.

Today, just before noon, his secretary ushered me into his office. He rose and offered me his hand. He was as polished as this room, in which he so obviously belonged; totally at ease, sophisticated, slender, and poised, he motioned me into a chair, then sat back. As we made the usual introductory comments people always seem to

225

make, he seemed to be completely immersed in what I was saying.

Brad joined us, and we lunched in the employees' dining room. Again, it was far more elegant than what I was used to at the Central Park cafeteria. Here I could even have a drink with lunch. It was not until near the end of the luncheon, when Mr. Conway saw some of my drawings, and he said, almost casually, "Why don't you get together with Brad and arrange to go over there when she's going to be outside?" that I could relax.

Then, suddenly it seemed, Brad and I were sitting in the golf cart, waiting. The door of the hospital opened, and Caroline was framed in the darkness of the doorway for a moment or so. She stepped out into the sunlight. Patty was bundled in her arms. She was all dressed up in pink and wrapped in a blanket, and there was no longer a sling on her arm. I waved at them eagerly, and Caroline waved back, delighted. "Hi!" She smiled. "Patty, look who's here!" She climbed up behind me and we were off. Suddenly I felt a little gorilla hand lightly touch my hair. I turned to see Patty's two bright eyes looking into mine. Her hair shone in the sunlight.

"Hi, Patty," I breathed. And touched her.

From the other direction another golf cart came sailing down the paths. Sherry King waved at me gaily as it pulled up. She climbed out of the vehicle, laden down on one side with Hodari, who clung to her, and on the other by heavy purses. The nipple of a baby bottle peeped out one side, and Pampers from the other. As she introduced me to a second woman, Alice Kipper, now Mopey's surrogate mother, Mopey reached out to pull at the bottle in Sherry's purse, and Hodari shrank deeper into her arms.

In no time we were surrounded.

"Hey, look at the monkey."

"Hey, lady, can I touch it?"

The crowd began to press in on us, and I found myself being led through the people down some steps and into the gray corridor that wound through the depths of the Ape House. Then, finally, we were led up and out again, to emerge into the "moat." We were center stage in the brilliant sunlight.

Caroline held Patty up for the crowd to see. When she put her

226

down, Patty sat quite still, one hand resting on her knee. When Caroline tried to move away a foot or so, Patty whimpered and looked piteously up into her face, grasping for her. When Caroline got up to get a bottle for her, Patty followed, crying, her little rear dragging on the ground in abject fear. She scooted fearfully past Mopey once with such a meek, frightened little action that my heart went out to her. The only time that I could remember any show of this timidity was the day of the accident.

"Don't worry, sweetheart, I'm here," Caroline cooed. "I wasn't going anywhere." She reached down to pick up the trembling baby, and Patty clung tightly. "Here, sweetheart, here's your bottle." Patty clutched at it, holding it tightly against her chest. Gradually the fear in her eyes softened again, and she relaxed, willing to sit next to her "mother," reassured by her presence.

"It's only the third time out," Caroline explained to me apologetically. "She's not quite used to it yet, are you, my little baby?"

I was disappointed. I had received the impression that the three little gorillas played together spontaneously. I had envisioned three little happy playmates grabbing at one toy or pulling at another, giggling together over some little chase or a game of hide-and-seek. Brad had told me that Mopey had tried to kiss Patty and had maneuvered her into position so that he could reach her face. And she had backed off, barking at him, not frightened, simply telling him he wasn't wanted. But I saw nothing of that here. Each of the women held her respective gorilla, separate little islands on the expanse of the stage. Over there Sherry coaxed Hodari to forget watching Mopey long enough to play peekaboo through her fingers, and over there, Caroline held Patty up so that a curious, demanding crowd would be satisfied.

But for the most part the babies sucked at their bottles of apple juice or had their diapers changed, were cooed over, and generally protected from hurricane Mopey, who lashed heedlessly across the moat.

Once Brad left for less amusing duties, Alice, of course, went into action. She never stopped running. Mopey would never be placid or tranquil; the only time there was even a semblance of quiet was when he was given his bottle. Even then his eyes shifted from person to person and gorilla to gorilla. What on earth was going on in his mind?

His primary form of entertainment was the pursuit of and attack of anyone, but particularly Hodari. And Hodari reacted true to form; he squeaked and squealed and made all sorts of frightened little animal sounds, burrowing deep into Sherry's lap. Mopey was whisked away, half carried, half flung in the opposite direction. Where did Alice get the energy? I thought as, breathless and panting, she hauled him off my back for the fourth time. He escaped and lunged at me again, his hands scrawling over the drawings. He scooted behind me and climbed up onto my shoulder to pull my hair or bite at my arms. Alice swooped down and pulled him off. He struggled, grabbed at my sack of supplies, and pencils went flying. The loose papers floated out in the breeze and skimmed over the moat. He bounded after them, snatching at this one and grabbing at that. Alice went rushing after him, nearly tripping over the box of Pampers which he overturned in his mad rush. On his way back he made a swipe at them and pulled the diapers out of the box. Across the cement field, he ran, a diaper flourished in hand.

"Oh, Mopey!" Alice wailed again, chasing after him.

All the cries of "Oh, Mopey" that filled the air did nothing to dispel the absolute bliss in which I found myself. Mopey bounded, Alice chased, and I sat in the midst of this wonderful hubbub with the two other little islands, calm and peaceful nearby. But it was when Caroline brought Patty Cake toward me and my sweet little gorilla padded curiously toward the papers on my lap that I held my breath. She reached out and put one finger on my cheek. She stroked my face. That was when I hugged her. There was nothing else in the world that I would rather have done, no place that I would rather have been! And she sat down by me and leaned against my leg, one hand resting on my knee.

"Well, it's time to go." Brad had mysteriously reappeared. Damn! It was over.

It was three o'clock, and the air was already beginning to chill. Shadows came over the moat. It was time to pack up and go home. Patty would head back to the hospital nursery, and the boys would head back to their playroom. I would head home. Reluctantly I packed my things. Soon the barrels would be filled, the boxes packed and the blankets taken up. The moat would be cleared again. We again filed down into the Ape House, into the corridor, past the kitchen and the familiar barred cages, then up again into the outer world. There everything was normal and as it had been a couple of hours ago. I glanced over the wall. The moat was empty of life. It looked naked. The crowd began to come toward us.

"Hey, lady, can I touch it?" The little girl's voice was shy with wonder.

And then the women, gorilla babies straddling their hips, laden down with purses and Pampers, climbed back into the golf cart and were off, waving brightly. They skimmed down the winding paths until they disappeared around a curve. I watched until they were out of sight. And wondered, hoped. . . . "Would there be a next time?"

CHAPTER 33
Patty and the Boys
Eighth Month at the Bronx Zoo

There were to be several next times, and each was nicer than the last. Every time I saw Patty she seemed less fearful. The sad, woebegone look no longer swept her eyes as if her mind were fastened on inner horrors. Her eyes were shiny and wide again; eagerness and curiosity were growing again. Of course, I was not able to see her very often, so each change was marked. It was as if some coil inside of her were unraveling slowly until it was all smoothed out.

At the moat she was always a bit uncomfortable and quiet, but at the playroom where I often saw her now, she was most at ease. The playroom was divided into two fenced-off cages, one of which was shared by Hodari and Patty Cake. The other Mopey shared with his surrogate mother, Alice, who slept there. During the day, however, it was his exclusively. The playroom was a child's dream, filled with all sorts of wondrous toys.

"Did Patty have bars or cyclone fencing?" Caroline asked me the first time I visited there. "Patty seemed so happy to see the bars on the jungle gym, as if she recognized them. Her eyes lit up, and she went running right to them. She still won't even try to climb the cyclone wire."

I looked at Patty Cake. She was climbing in and out of a little red wagon, running merrily in and out of a serpentine blue canvas tunnel, and running after a ball. This was the first time I could really see how active she had become in the last weeks. She was happier than I had seen her since the accident and was even walking on her knuckles like a real little gorilla, something she had never done before. Now and again she even made soft, feminine charges across the carpeted floor, enjoying the extra spurt of energy they held. She ate eagerly, enjoying each mouthful of baby food. It was seldom that any fruit was strewn across the floor, but when it was, she approached it curiously, and as she had in Central Park, she bent down to sniff it before picking it up and looking at it closely. And it stirred something inside me.

No one could say that these animals were deprived! They had everything that money could buy a baby gorilla. Yet I felt a tugging sorrow. No matter how much the people cared for their welfare, Mopey and Hodari would never have the experience of growing up as a member of a gorilla family in these early months. At Central Park the social, not the physical, environment had always seemed the more important to Kongo, Lulu, and Patty Cake. The lack of toys and outside stimuli seemed almost trivial, compared with the kinds of care and activity that they shared so naturally and the control they had, in part, over their own lives.

There seemed to be a very big difference in the ways that the women at the Bronz Zoo treated the little gorillas and the way Lulu had handled her daughter. The women seemed to adore their charges and certainly did everything they could for them. Yet the ways they handled the babies seemed clumsy and unsure, forced. They put human demands on them and projected human fears as well. They made so much of an effort to satisfy the gorillas in this controlled environment that it seemed to become an artificial existence for them. With Lulu everything had seemed so easy and unstrained. At home Patty had been part of the situation, not the object for which the situation was created. The thought struck me hard. She might never have that again.

I shook the thought away. Patty was merrily chasing a bright-red ball, and Hodari was standing there watching, wide-eyed, open-mouthed, and giggling with vicarious glee.

Hodari was not afraid of Patty Cake, as he was of Mopey. He must have recognized her gentleness. Although the two of them often occupied the same cage, it was not often that they actually played together; Hodari was not yet up to that, and Patty was content to play with her toys by herself. But Hodari did begin to venture out on his own, and finally in May he had some small cowardly overtures. Still shy and unsure, he could at least begin to enjoy himself. Once, as Patty was scrambling through the tunnel, her shape bulging out against the material, Hodari became playfully adventurous. It was unlike him. Tittering and giggling in anticipation, he followed the bump that traveled down the length of the tunnel and then, with daring impetuosity, he swiped out at the giggling protuberance as it wiggled by. Then he ran, exultant and frolicsome, his eyes alight with a mischief he had never revealed before. Patty emerged from the tunnel and ran after him, clucking. Finally, she had an animal friend. But when she reached out to touch him, he backed away and cowered. His friendship would be a distant one.

One day, as I arrived, the women were aglow with news. Hodari had stood up to Mopey as Mopey had charged, on the other side of the cyclone fence, of course. Hodari had then scrambled away, so pleased with himself that he couldn't stop giggling for minutes. It was a momentous occasion.

The days I spent at the playroom were cheerful and busy, Patty swinging high above the ground or running happily from one toy to another, Mopey blustering, and Hodari timidly but overtly enjoying the action around him. The women sat on the outskirts, preparing food, feeding, diapering, and, when necessary, protecting their own charges.

Then came a day in which the press was finally invited to view Patty at the Bronx Zoo. They arrived en masse.

It was a madhouse as we made our way through the crowds of reporters and cameramen at the moat. The women left me with the crowds as they entered the Ape House to reappear inside the enclosure.

They had barely had time to settle down with the babies when the questions and directions began; the noise and confusion increased.

Whether it was the memory of that other day when the cameras had pointed at her relentlessly as she lay helpless and confused on

233

a hospital table or it was simply the excitement and hubbub, Patty became more and more disturbed. She began to clutch Caroline as fear grew in her eyes. Her stool became loose and began to trickle out of her diapers. Then it began to flow over the cement.

Caroline tried to comfort her, simultaneously trying to call answers to the press, holding Patty in position for a photograph and gently soothing her. But Patty began to cry; feces poured out of her, she cowared against her surrogate mother, clutching at her frantically. Finally, Caroline simply stopped complying with the press' requests and turned her full attention to Patty. When, at last, the interviews were over and the people drifted away, it took another hour before Patty calmed down and timidly ventured out of Caroline's lap. When Caroline moved, Patty whirled and clutched at her until she was sure she was not alone, not deserted. Slowly the fear dissipated. She would soon be scampering along and reaching up to the horizontal bar of the jungle gym, her cheerful self. But I knew now that the fear was still there. She had not forgotten terror.

I visited Patty at the Bronx Zoo for the last time on May 10. We spent the morning at the playroom as usual. Then, as Alice and Sherry got their charges ready for an afternoon at the moat, Caroline and I took Patty into the backyard for a picnic.

We sat under a grape arbor that Sherry was growing. It had rained that morning, but now the sun was shining. The sky was pure, deep blue, and the grass was spring green, covered with dandelions. As we unwrapped our sandwiches Patty sat on Caroline's lap and reached out to finger the grass.

"Why don't you put her down?"

"I never put her on the grass before. Don't you think it's a little damp?"

"Oh, Caroline, she'll love it. Come on."

Patty clutched some grass in her fist and yanked a clump out of the earth. She looked at it, sniffed it, then put it to her mouth. Caroline took it away with a firm no.

Patty looked beautiful against the green; her hair shone as she bent over to scrutinize a weed. She rolled on the grass, stretching out over and over again. She plucked a dandelion, and holding it by

234

its stem, she climbed into Caroline's lap to show her what she had found.

Later, at the moat, the babies were introduced to a plastic. wading pool. It was still too chilly for an outdoor bath, and the pool was empty. The animals, one by one, were placed in the tub along with a toy or two, just to get used to it. The afternoon was peaceful and calm. And sun-soaked, it slipped away.

At three thirty the babies nodded and began to drift off to sleep. Dozing, they were carried off to their golf cart. Hodari's head rested on Sherry's shoulder; his arm was limp down her back. And Patty, cradled in Caroline's arms, was snug and secure. Mopey, for once, was docile, but as they drove away, one eye opened, and he reached for Alice's bag.

It was the last time I was allowed to see them. A few days later I called, as usual, and spoke with Brad. He sounded very apologetic, and I knew what was coming.

"I wasn't there," he told me, "but I understand that some children somehow got into the nursery with Patty. They're not supposed to, but they did. I think their mothers work here somewhere. Anyway, one of the mothers called to tell us that her child came down with the mumps. It was very thoughtful of her, don't you think?"

Patty was in quarantine.

CHAPTER 34
The Controversy
Eighth and Ninth Months, Central Park Zoo

Far away from the toys, from Lulu and Kongo, from Mopey and Hodari, a controversy was raging over Patty Cake's future. It took place in a world in which men sat behind desks and argued her fate; where letters were sent back and forth between executive and administrator; where the get-well cards which Patty received (35,000 of them) were counted and re-counted, and where which zoo received the greater number was an important issue. A group of protesters marched up and down Fifth Avenue above the Central Park Zoo, decrying its condition. "It's too old!" they cried. "It's not even a zoo! It's a nineteenth-century menagerie!" Letters to the editor were written to the New York *Times* and the *Post* and printed.

Fitz, the keepers, Andrée, and I watched the going-ons with angry, sad eyes. The zoo *was* old. It dated back to the mid-1800's. The cages were barred and bare. The space was often inadequate.

Fitz had done everything he could to change it. When the birds refused to mate in their own naked cages and he had asked for help, he had been given $600, which he spent on plants. The rest he did himself: He hauled rocks from the site where the earth was being blasted for a new subway; he wandered around the park to get tree limbs and branches; he set pipes into the cracks between the

237

rocks and sent water streaming through them, creating pools of clear water. The birds began to mate.

But he could not create a new environment for the gorillas. Too much money was needed; too much unavailable expertise was required. The only physical change that he could make in anticipation of Patty's return was to set up a heavy plank across one of the outer cages, an extra dimension for the apes. It seemed very little.

We knew that many people would not understand that the care the keepers gave to the animals could not be any better, that the relationship among the gorillas was intimate and agreeable. "Their relationship must have meant something," we told each other. These animals were happy enough to have cared for Patty, brought her up, played together, and lived in some semblance of social normality, while in many other zoos gorillas were not even well enough adjusted to mate. In San Francisco young gorillas were even being shown pornographic movies in the hope that they would learn through example what they had not been able to observe in a normal way. But it didn't seem to be helping them.

We felt depressed and helpless. The long list of Patty's ailments which were repeated over and over again by the Bronx Zoo staff seemed to point a long and almost accusative finger at Central Park.

When I met Mr. Conway for lunch at the end of April, he had enumerated the problems and had sounded so convincing that if I had not been familiar with the situation and the animals, I might have agreed with him that Patty should remain in the Bronx.

"Our interest in Patty is several fold," he had said. He leaned back in his chair and smiled kindly at me. Brad House sat forward, listening. "Not only because she's a unique animal, the first admitted"—he emphasized the word—"gorilla that has been born in Gotham, but rather because she is a member of what is an increasingly endangered species. And she's a female member." He paused as we were served our lunch, then continued. "And although Gloria Steinem hasn't pointed it out yet, the fact is that in matters of wildlife, females are a lot more important than males for the good old biological reason that one male can impregnate a lot of females, but the females cannot impregnate the males. So that the number of females determines the number of young that they are going to have."

238

He looked at me as if to ask if I had so far followed this logic, then went on decisively.

"So Patty Cake is rather special. She's a captive-bred gorilla, she's a female, and the chances are that very, very few gorillas will ever be imported into the United States again. We're going to have to try very hard to breed those animals we have. . . ."

Patty Cake was in real trouble!

"Frankly, the best thing that could have happened to her was a broken arm." I must have looked jarringly disbelieving, for Mr. Conway leaned forward in a convincing, more businesslike posture. "To give you an idea, Patty Cake, when she came to us, weighed eight pounds. Now gorilla babies at seven months of age should weigh between fourteen and sixteen pounds. She was in an advanced stage of malnutrition and half her normal size.

"We think that Patty's arm may reflect some decalcification. Certainly her gums were in a very poor condition, and generally she was at a very low level." He lowered his voice to a somber tone. "She was also carrying," he informed me, "a load of three different species of parasites"—but then his voice took on a relieved and cheerfully brisk inflection—"which we have since cleared up. However, if she had stayed with her parents, because that female was, as Brad has characterized her, an irresponsible wench, I think the baby would have been lost."

Instantaneous protest must have shown on my face, for he was quick to reassure me. "Now Lulu shouldn't be blamed perhaps. . . ."

Brad stirred and interposed, "No, I think she'll settle down now." The two men began to reassure me of Lulu's future maternal capabilities.

"So," Mr. Conway concluded, "the fortunate thing is that she's alive. As of last Monday she weighed eleven pounds which, is an increase of thirty-seven percent over her original weight when we received her—that's a pretty good weight gain. I hate to think what I'd look like if I gained thirty-seven percent of my weight!" We laughed at the joke, and his voice became warm. "She's coming around beautifully now."

Again Brad nodded emphatically in agreement.

Mr. Conway continued. "Now there's another problem. With any of the great apes it is not practical, as you will see, to introduce her to her parents. In the first place, it is dubious that they would recognize her as their baby. They would not know her name. . . ."

His cool, smooth voice went on and on, and his words blended into one distinct message: Patty Cake should remain at the Bronx Zoo. She had been on the verge of death, had been rescued only by the efforts of those people at the Bronx Zoo. The Central Park Zoo staff not only had been lax in their treatment of her, but had actually caused her sickly and deficient condition, and if they were really conscientiously concerned for her health and survival, they could not possibly wish to return her to those conditions. Lulu would not accept her, and if she were returned, she would end up alone in a cage, dealt a life of unnatural solitude.

"We don't think that there's been permanent brain damage," Mr. Conway reassured me, and Brad again hastened to agree. Mr. Conway smiled at me. I looked down at my plate.

So much of what he said was indisputable. Gorillas are an endangered species, and it is difficult to breed them in captivity, although that trend had already turned. However, only a few days before that luncheon M'wasi, a female gorilla at the Bronx Zoo, had had a cesarean performed on her too late. The baby, too large for the mother's birth canal, was dead. And in Barcelona Snowflake's little baby had died of pneumonia after spending all his young life in an incubator.

The parasites did exist; most zoo animals have them and are dewormed on a regular basis. Patty was small and had gained weight. As I sat there listening, there was little I could say.

The press conference of May 7 which the Bronx Zoo had finally permitted had not helped the situation. Patty Cake appeared on television healthy and more robust than she had ever been before, playing in the sunshine, playing "with" the boys, while Dr. Dolensek reluctantly spoke of her previously weakened condition. The keepers at Central Park were morose, for what was stated then seemed to be a direct attack on their abilities to care for Patty Cake.

A day or so later the keeper Eddie Rodriguez nearly cried out his hurt and frustration in front of the television cameras, then hoped that the public would rush to the defense of Central Park and de-

mand Patty's return. But I knew that it really didn't matter. Whatever happened would be decided by the men who sat behind the
desks, and with the press conference our hopes sank.

We all were depressed. An old rumor was rekindled: The Bronx
Zoo really wanted to take over the Central Park Zoo altogether.
With the accusations and the controversy, Patty seemed to be
caught up in that old struggle as well. The power that we believed
the Bronx Zoo wielded threatened us. The Central Park Zoo
seemed helpless. What chance could there be of getting Patty
back?

Suddenly other rumors began to circulate, rumors that made us
think that perhaps Central Park had not lost Patty. Parks Commissioner Richard Clurman was thinking of hiring a consultant. A little
hopeful now, we went about our routines, while the decisive
moves were taking place behind closed doors.

Commissioner Clurman had no personal interest in gorillas. Except to think about generally improving the zoo, he had never even
thought much about them until the day that Patty had been hurt.
Then he became totally involved with all the decisions that affected
her. No one ever saw him cuddle her or pose with her, but in his
office he knew exactly what was happening. It was he who had
made the decision to leave her at the Bronx Zoo originally. Now,
as the controversy became more and more heated, being called a
custody battle by the press, he would decide where Patty's permanent home would be.

I have always thought it would have been easier for him if he had
simply left her in the Bronx. It was inarguably better equipped, far
more elaborate, and it was certainly more attractive than the Central Park Zoo, and its personnel were better trained. In very little
time the public would have forgotten all about the controversy. Or
if Commissioner Clurman had been determined from the beginning
to bring Patty back to Central Park, as many other people might
have done (certainly wanted to do), it would have been easy to use
his position in government almost as a threat against the Bronx
Zoo. The city provided it with funds; it was city land on which the
Bronx Zoo stood. Instead, he chose to do what was right for the
animal.

He had no idea what was right for her. There were plenty of "experts" around him, but they all had different opinions. And no doubt, some of those opinions were, to some degree, shaded by either a competition between the zoos or a personal involvement with the animal. The commissioner knew that he certainly had no expertise on the subject and was totally unqualified to make any kind of knowledgeable judgment himself. Therefore, he asked his administrative assistant, Dr. Donald Simon, to find a consultant whose expertise would be unquestionable and whose impartiality so absolute that each zoo would accept him as an authority.

After much inquiry and research, every road that Donald took led him to the Yerkes Regional Primate Research Center in Atlanta, Georgia, and to Dr. Ronald Nadler.

Dr. Nadler seemed the logical choice. He was highly recommended by such people as Dr. Leonard Rosenblum of Downstate Medical College. He had a PhD in psychology and additional training in neuroanatomy, neuroendocrinology, and neurophysiology. He had published more than two dozen articles, held numerous honors, fellowships, and memberships, and was at present a developmental biologist, involved in a comparative-developmental program with the great apes at Yerkes, probably the most prestigious primate institute in the United States. It was an impressive array of accomplishments, and the problem at hand seemed to be right up his alley.

It was only after his name had been submitted to the commissioner and approved by him that Commissioner Clurman called Mr. Conway and told him what he was prepared to do, and in a short time Mr. Conway agreed that the choice of Ron Nadler was a good one. It was then that Donald called Ron.

The phone call did not come as a complete surprise to Ron. Dr. Leonard Rosenblum had mentioned to him that his name was under consideration, and he had, of course, heard of Patty Cake. Everyone had. But the situation, as it was described to him by Donald, seemed to do with the physical properties of the zoos and the physical health of the animal. It seemed distant from his own interests. He declined the job and referred Donald to Dr. Michale E. Keeling, the veterinarian at Yerkes.

But when he hung up the phone, second thoughts began to disturb him. He felt a nagging annoyance with himself. He reached for his pipe, lit it, and watched the smoke curl up. The more he

242

thought about it, the more disturbed he became. Perhaps the situation was not so cut-and-dry as he made it out to be. Perhaps he had just turned down an opportunity to involve himself with a problem having to do with reproduction in gorillas and with mother/infant relationships, a subject in which he was very interested. He sighed and resignedly went to tell Dr. Keeling what had just occurred. When, much to his surprise, Dr. Keeling told him that he was not interested in the problem, it was Ron's second chance.

He arrived in New York on May 15. The next morning he borrowed a friend's car and told his friend that he would be back in a couple of hours. He thought that he would meet the curator of mammals at the Central Park Zoo, have a brief conversation with him and, on the basis of whatever information he was given, make a judgment then and there. Certainly he did not expect to be met at the parking lot and then graciously escorted to the office of the commissioner of parks. With a nervous growing awareness that this was something beyond his initial understanding of the matter, he was ushered into the imposing formal conference room, where Donald formally introduced him to Commissioner Clurman and then to Deirdre Carmody of the *Times*. Seated at the long conference table, overseen by portions of George and Martha Washington in gilt frames, lit by a stately brass chandelier, he was realizing by the minute with growing excitement that this was a serious and highly publicized event. He was almost a celebrity; people were aware that he was coming, they were even aware of his $100-a-day-plus-expenses fee, and were already discussing his credentials. An unobtrusive tape recorder was silently taking down every word. And a reporter was scribbling additional notes. He began to reevaluate the situation. He had, when he agreed to act as consultant, thought of the job as a relatively insignificant one, interesting, perhaps to those people involved and to himself on a professional basis. It was obviously of far greater magnitude.

Commissioner Clurman was saying, "Let me ask you a Watergate-type question." He paused. "Do you have any relationship with the New York Zoological Society or the Central Park Zoo that makes you less than an impartial observer?"

Ron looked into the commissioner's eyes and shook his head negatively.

The meeting had begun.

243

Mr. Clurman outlined the situation for Ron and explained the reasons for asking him to act as a consultant. As he spoke, it became clear that the kind of investigation he expected called for extensive and intensive research and consultation. It would take days, not hours, fully and comprehensively to interview the people who were in any way involved with the animal, to study the physical conditions of the two environments, to go over medical reports. Anything and everything would be available to him at either zoo. The Central Park Zoo would assign a public relations representative, Sandy Lipshitz, as an escort, to make certain that he had everything he needed. She would accompany him to the Bronx Zoo, as would Dierdre Carmody.

There were several questions to be considered, Mr. Clurman emphasized: whether or not Patty Cake should return to the Central Park Zoo and, if the answer was no, whether or not she should remain at the Bronz Zoo.

"One other thing," the commissioner concluded, "when you have made your decision, please send the report directly to me. And in no way indicate what your decision is to be to anyone else. Nothing will be published or disclosed until this whole thing is over with. . . . I would like to know what is happening along the way and be in touch. I expect to be fully involved."

It was crystal clear: Dr. Nadler would make the recommendation. Mr. Clurman would make the decision.

Of course, I knew nothing of any of this. I was still in the Lion House, taking notes and drawing and waiting anxiously for the appearance of the expert. I envisioned an impressive figure, an elderly man with a beard, carrying tattered loose-leaf notebooks and wearing a tie haphazardly tied and long since askew. I was right about the beard. When Fitz strolled into the Lion House accompanied by Sandy and an immaculately groomed, fashionably attired young man, I quickly altered my ideas. This had to be he. He was engrossed in Fitz's conversation. They stopped in front of the gorillas' cage where the accident had occurred, and I strained to hear what they were saying.

"This is . . . uh . . . Susan Green," Fitz was saying. "If anyone can tell you about the gorillas, she can. She's been watching them and taking notes every day since September."

244

Ron looked into my eyes, intently curious, and I was off, eager, more than willing to pour out description after description of the relationship between the mother and child, the father and mother, the father and baby now that I had the chance. He listened to what I said, and the more I spoke, the more fully aware he was that I was not a professional primatologist or psychologist. As I rattled on and on, jumping from one lovely memory to another, he wondered to what degree he could rely on my statements. Over and over again he said, "I don't want your impressions. I want the factual information. What did you see? How many times did the baby nurse? How did the mother carry her? Were you ever afraid?" he asked me. "Did the gorillas ever make you feel fearful?"

I thought back. Only twice that I could remember: the time they had their first fight and Patty seemed to be in the midst of the melee and the time Patty had fallen from the crossbar and lain so still. But, I told him, I could go back over my notes and sift things out if he wanted.

"How thorough are your notes?"

"Well, I just took down what I saw. There are about five hundred and fifty pages now."

"Five hundred and fifty?" He looked astounded.

"Single-spaced," I hurried to reassure him.

He said that he would be interested in seeing certain sections of them: those concerned with Patty's eating habits, nursing habits, and the relationship among the animals, particularly in the two or three weeks before the accident.

We had been talking for about an hour when Fitz suddenly cleared his throat and explained that Dr. Nadler was expected at a luncheon. "Would you like to come?" he asked me.

Would I?

We continued the conversation as official Parks Department cars picked us up and drove us to the restaurant, a beautiful place in the middle of the park. This, too, was a new experience. I had never been in a chauffeured car before, or swept into a restaurant where the maître d' had escorted us with such an air of importance to our table in the sunshine. I almost wished that I were wearing something lovely and floating instead of my zoo-green slacks and sweater that smelled distinctly of gorilla.

But it was difficult to enjoy the meal, for even in this lovely and abundant setting I was aware of the tension that prevailed and felt a desperate need for us to describe Lulu's concern and care for her daughter, a need for us to convince Ron of the wonderful and unique relationship that the animals had held. We knew that up at the Bronx Zoo Ron would see surroundings far superior to those of Central Park. How could we explain to him what those animals had had?

It was with anxious frustration that, when the luncheon ended, I watched him again get into the chauffeured car and with Deirdre and Sandy start the drive up to the Bronx.

Ron and I met again the following day. There was no indication of what had transpired at the Bronx. I was sure that he had had a cordial welcome, and I was desperately curious. But neither one of us mentioned it. I had brought along all my notes and had photocopied those for which he had specifically asked. We simply went over detail after detail again and again. By this time he was fully aware of the situation and had formulated the method by which he would handle it. His questions were even more specific than they had been the day before. He knew that the possibility of greatest danger to Patty Cake lay in reuniting her with her parents. The easiest thing, and the safest, would be to recommend that she remain in the Bronx. But he had nagging doubts. He was not sure how it would affect her future sexual capabilities. As Mr. Conway had stated, she was a female member of an endangered species. Ron had found that often the influence that people have on the primates that are hand-raised suggests that the animals are better off if left only with the members of their own species with as little interference from humans as is possible. Too, he could not discount the possibility that the people at the Central Park Zoo had colored the relationship and had projected a rosier or more congenial environment than had actually existed. And if the relationship had been as they described it, was the past a good prediction for the future? He kept digging.

We kept going over the same notes and the same questions. By the time he was ready to leave New York Ron felt that he had gleaned as much information from all of us as he possibly could. But as he hurried up the steps to Fifth Avenue, rushing to catch a plane back to Georgia, I stood there wishing that there was something else I could tell him. Had I forgotten anything? What had he been told at the Bronx?

He had said that there was a great deal of literature on the subject of mother/infant relationships of primates that he wanted to read and people whom he wanted to consult. It would be at least a week or ten days before he would give his recommendations.

I went back to Lulu and Kongo.

More than a week passed.

Ron came to his conclusion. As agreed, he first got in touch with Commissioner Clurman. When the commissioner read the report, there was no doubt in his mind about what to do. But how to keep it secret? Emotions were already running so high over Patty's future that the commissioner wanted no further discussion or argument over his decision. If Mr. Conway agreed, they would announce at a press conference at which both parties would agree to the decision publicly. Until then no one else would know.

As it happened, the commissioner was supposed to drive up to Wave Hill, a beautiful old estate which overlooks the Hudson River from the northern Bronx. It is not far from the Bronx Zoo. Mr. Clurman called Mr. Conway and told him that the report had arrived and that he had made a judgment on the matter. "I would like to send my car for you," the commissioner said. "The report will be in the car. If you can read it on the way over here, then we can decide what to do about it when you arrive."

Mr. Conway asked the commissioner what the report contained.

"I would rather you read it first," Mr. Clurman said.

No one knew that they met; no one even knew that the report had been received.

Ten days had passed since Ron had hurried away to catch his plane. Ten days, and there was still no sign of the report and no word when it might arrive. Each day I went to the zoo expecting to hear something regarding Patty's future, but there were only the usual rumors and speculation. As I worked, my mind was often preoccupied with the problem. And of course, I brought it home with me.

247

"No," my husband told me, "it's going to work out. If he had made the easy decision, you'd know by now."

We were sitting in the living room that June 1, trying to decide whether we could risk going away for the weekend. The news had to come soon, and I wanted to be there when it did. We were in the middle of the discussion when the telephone rang. My husband answered.

"Sue," he called, "it's Donald Simon of the Parks Department!"

I jumped from the couch and almost fell over. Suddenly and surprisingly my heart lurched and my knees seemed weak. I had thought I was so calm.

As I took, the phone my husband stood by, watching my face.

"Hi, Donald."

"Hi!" His voice was cheerful. "Do you think you can be at the Bronx for a press conference Tuesday?"

I sat down. "What is it?"

"I can't tell anyone yet. We can't let it out until the conference. All I can tell you is that we think you ought to be there. It's at one thirty."

"Of course . . . sure, I'll be there."

"Good, I thought you'd make it."

We spoke for a few more moments; then I hung up and looked up at my husband. "What do you think?" I asked breathlessly.

He was smiling. "I'll bet she's coming back." He grinned. "The Parks Department would not be calling a press conference, and I doubt that the Bronx Zoo would be calling to invite you if she were staying there." I looked at him hopefully. "Don't worry, she's coming back," he reassured me.

We went away for the weekend.

CHAPTER 35
The Decision
June 5, 1973

It was swelteringly hot in the crowded car that pulled out of the Central Park parking lot on its way to the Bronx Zoo. It was not until now, when we were actually under way, that Donald took some papers out of his briefcase and handed a number of them to me. His eyes were filled with a twinkling excitement as he leaned over the seat toward me.

"I think you'll be interested in this," he said.

It was Dr. Nadler's report.

> A recommendation is made to return Patty Cake to her parents at the Central Park Zoo.

There it was! Now I began to smile.

> The recommendation is based on the judgment that an infant gorilla is more likely to develop into a socially competent and reproductively adequate animal if it is raised in the company of its parents as opposed to being raised with a group of peers. Conditions requiring attentions and suggestions for implementing the reintroduction of Patty Cake and for follow-up evaluations of her health and welfare are included.

It was quiet in the car. I continued reading.

> The major source of information regarding the quality of behav-
> ioral interactions between Patty Cake and her parents was Ms. Su-
> san Green, a consistent and conscientious observer of the gorilla trio
> from a time shortly after the birth to the time of separation.

Again my heart jumped, and I looked up, staring into space for a
moment, drinking in the idea that I was, in part, responsible for Dr.
Nadler's decision. It was a little nerve-racking and a bit frightening. My dreams had become a reality. And there was no way to tell
what the results would be. What if I were completely wrong? I did
not want to think about it and looked back at the report.

> Although one cannot rule out the possibility of bias on the part
> of those who portrayed the favorable impression of Lulu's maternal-
> ism, the detailed verbal descriptions and notes depicting the mother-
> infant relationship, as distinct from their interpretations, were so
> thoroughgoing and complete as to leave little doubt as to the general
> tenor of the events of interest.

I was glad that I had been there, glad that purely by chance I had
recorded my observations, glad that I had been able to tell Dr. Nadler what I had witnessed. I could feel my eyes filling. I glanced
around me, but Donald did not notice my slip into unprofessional
reaction. I was able to enjoy it and dream about Patty Cake's return. Perhaps today someone would hand her to me, and I could
hold her, carry her to a waiting car and home.

Ours was the first car to arrive at the Bronx Zoo. We located the
building where the press conference was to take place, but there
was not an official in sight, only a class of schoolchildren who were
watching the antics of the sea lions nearly. We felt awkward, conspicuous, and a bit ridiculous as we stood perspiring, shifting, and
shuffling, waiting for someone to appear.

Finally, Simon Dresner, of the Bronx Zoo Public Relations Department, arrived. But with horrendous, devastating news: Not far
from the zoo, at the 174th Street subway station, a transit police-
man had been shot and killed. Most of the press had gone there.

The two policemen who had stood joking with us a moment ago blanched, leaped onto their scooters, and were off.

A pall settled over us. For a moment even the dreadful heat seemed to lessen. After the exclamations of horror the conversation, such as it was, ceased altogether.

The day grew hotter, and slowly each of us, subdued and lonely, wandered into the huge room where the conference was to take place. Once inside, little by little conversations began again. The sound of voices rose from a hesitant, self-conscious murmur to normal tones. Commissioner Clurman arrived, and the press began to straggle in and set up their equipment. I saw some familiar faces; Dierdre Carmody waved at me across the room, smiling, and Peter Coutros of the *Daily News* came by to say hello. It was becoming noisier and livelier as the rest of the press came and began to go about their usual business: setting up, interviewing, photographing, rehearsing. Finally, I saw Mr. Conway and Brad hurry, businesslike, into the now-crowded room. I found a chair against a wall, far from the podium, and sank into it, took out my notebook, and wiped my forehead. It was hotter in here than outside. Someone opened a door, and a whiff of a breeze titillated us with it's sudden life.

Ron came toward me, smiling nervously. We shook hands, and as he sat down, he leaned toward me intently.

"I sure as hell hope you told me the truth," he murmured, half-laughingly.

The nervousness gripped me again. What had I told him? I could not remember.

Suddenly the room was illuminated by the glare of lights. The murmurings of the people ceased, and Commissioner Clurman walked toward the podium. Ron took his seat at the front of the circle of chairs. As I watched him walk away from me, I realized just how important this decision could be to him personally. If he were right, then the events which came as a result of his decision could be revolutionary. This would be the first time that a reunion between a mother and infant gorilla, separated for such a long duration, occurred. There had been experiments under highly controlled circumstances in which monkeys had been deliberately

251

removed and then returned to their mothers. And of course, other gorillas had been introduced to groups. But they were much older than Patty Cake, and the situations were different. If this reunion were successful, it could establish a precedent which could be applied to other situations. It could establish the acceptance, among people who still believed that this was a facility reserved only for humans, of the existence of an emotional memory in the gorilla.* If the reunion were a failure, Ron would be considered responsible. My own part in this event made me nervous. There were so many responsible people who believed that I was wrong—Mr. Conway, Brad, Dr. Dolensek. If it failed, would I, too, be responsible?

Commissioner Clurman began to speak. He, too, had a share of responsibility in this event. From the beginning he had been determined to discover what was the best for the animal. He could easily have handed Patty over to the Bronx Zoo. There would have been no risk, no gamble, and the public would have soon forgotten the controversy. But that would not have satisfied his sense of responsibility to the Parks Department for which he worked, or to the animal. Now he stood before the microphones and publicly announced that Patty was going home.

Then he thanked the people at the Bronx Zoo who had taken such good care of her.

"How much is the bill?" one of the newspeople asked jokingly.

He replied that he hoped that the Bronx Zoo would understand the finances of the city of New York and be reasonable in its request for payment. About three weeks earlier, I had overheard two girls talking on the bus.

"I heard that Patty Cake's on Medicaid," one of them said.

"You think she's got a card? Blue Cross. Name—Patty Cake Gorilla." Everyone who had overhead had smiled or chuckled. Apparently it hadn't been true.

Dr. Nadler then took his place at the podium. First he stated his recommendation and his reasons. Then he and the commissioner

---

*I had only recently been approached by a woman who said that she represented a textbook company which was in the process of writing a chapter or two concerning gorillas. They wanted me to be included in the chapter. "But," she said, "I don't want to hear about any of this emotional crap they're handing out." I told her that I was sorry, but I thought she had the wrong person.

accepted questions from the press. Some of the questions angered me. They seemed to dwell on the horrendous things that could occur if the reunion failed. What would happen if Lulu rejected her? What would happen if Kongo interfered? The questions were unanswereable.

Then suddenly it was over. We began to gather our belongings and to swarm toward the door and fresh air.

Brad passed by. He bowed slightly, said, "Hello, Susan," and walked on. I was sorry to see him so formal and strained. He had truly believed that Patty belonged in the Bronx. He still did. Just as I believed that she belonged with her mother.

CHAPTER 36
At the Bronx Zoo with Patty Cake
June 7–10, 1973

The following day a few of us met to discuss our plans for Patty's return, to determine when the cages would be cleaned, when Lulu and Kongo would be wormed, who was to care for Patty and how and where. We could determine where the reunion would take place, if not when.

"Don't prolong all this preparation," Dr. Nadler had said. "The sooner she is reunited with her mother, the better. She has been away long enough."

He would return from Georgia for the reunion. In the meantime, it was up to Eddie, Ronnie and Fitz.

"Have you heard the news?" Someone stuck his head in the doorway.

Fitz shook his head. The keeper continued, his voice grave. "They found the baby gorilla at the Lincoln Park Zoo [Illinois] on the floor this morning. Her mother bit her. They said something about brain surgery." The head retreated, and the door closed.

Oh, God!

The following morning I once again opened the door to the veterinary hospital at the Bronx Zoo. This time I walked by Dr. Dolensek's office and straight down the hall without trepidation. I belonged there.

I peeked in the window. Ronnie was sitting on the floor in the middle of the room, Patty leaning casually against her leg. Eddie was happily watching from the sidelines. I was thrilled. We had expected Patty to be timid at this first meeting, yet there she was, leaning trustingly on Ronnie as Caroline sat in her usual chair by the playpen. Singing, I pushed opened the doors. Patty looked toward me expectantly.

"Hi, Patty!"

There was a flash of recognition. As I sat down on the floor, she came to me, sliding over the floor in her rush. She climbed up onto my lap, and I slipped my arms around her. She grabbed at my hair and pulled, then bent over to nibble affectionately on my hands. She raised her head, looked into my eyes, and giggled. Just for a moment I held her close.

Ronnie and I greeted each other happily. Then I looked up at Eddie. He was beaming, proud to have been the keeper chosen to come here to care for the little gorilla, happy to see her. Yet Patty had not gone to greet him the way she had Ronnie and myself. "She doesn't take to men very easily," Caroline explained apologetically. It seemed to be true, for although Patty would look up at him from the corners of her eyes as she sidled past him, she made no attempt to reach out and touch him.

Patty went about her usual business. Ronnie had brought her a present, a string of plastic ponies strung on elastic string, the kind of toy that is strung across the front of a baby carriage. Patty loved it. She pulled it along after her as she went from one thing to another. It caught around a table leg, then snapped, flying like a slingshot, when she tugged on it. It clattered on the floor as she scrambled from one of us to another and swung out behind her as she climbed. She would not let it go, even when she laughingly came to me, trying to grab my pencil, or spun around to rush at Ronnie, who sat on the floor, speaking quietly to her. She was having such a wonderful time that Caroline thought it the right moment to leave for the first time. The door closed behind her. Patty was aware that she had left. She looked toward the door, but she continued what she was doing, chasing the keys that Eddie was waving at her. She left him to come to me. She bent down to sniff my red umbrella, then to tug at it. She climbed onto me and bent over, stretching her mouth wide over my hands to pinch the skin with her teeth in gentle, and not so gentle, mock biting. She grabbed my purse and walked off with it, dragging it along on the floor behind her, together with her plastic ponies. She looked back over her shoulder, then quickened her pace, scooting away mischievously. "Hey, Patty . . . leave my purse here . . . hey, bring it back!" Some pencils and erasers spilled out onto the floor. And as I reached over to remove her new toys from her eager hands, she spied Eddie squatting on the floor nearby, dropped the purse, and headed toward him instead.

When Caroline returned, Patty hardly seemed to notice. So far, so good.

As long as she had company, she seemed to feel comfortable, Caroline told us, but when she was left alone, she began to scream. It seemed natural enough; gorilla infants are never naturally alone. No one turns out the light and closes the door at night. Rather, baby gorillas cling to their mothers and sleep nestled against their bodies. Their mothers are always within their view, and it is to them that they return to feed. Often other female gorillas keep track of them and play with them, like "aunts." Were Ronnie and I "aunts"? What would happen when it came feeding time and Caroline and I left the room? Ronnie was to give the baby gorilla her next meal.

Caroline had already begun to fade into the background, less and less active with Patty Cake. Ronnie came forward and began to replace her as Patty's surrogate mother. Now, as Ronnie began to prepare Patty's bottle, Caroline and I slipped away, sneaking out to the window to watch Patty's reactions.

"Up ya go, P.C.," Ronnie said amicably and offered Patty some of the baby food. But the little ape pushed the spoon away and began to squirm out of the woman's grasp. "OK, OK. You don't have to eat it. Hey, wait a minute, look what I've got for ya!" She offered Patty the bottle. But the gorilla did not want that either. She wiggled and squirmed, and Ronnie let her go. She was not going to force Patty to accept food from her. It was then that Patty Cake must have realized that Caroline had really left her in someone else's care, for she ran to the door. She stood looking up at it for a moment or two, then touched it with her hand. She tried to open it, pulling at it, then looked around at Ronnie, who still sat on the floor where she had left her. Patty vascillated between the door and Ronnie for a moment, trying to make up her mind. She chose Ronnie and went scurrying back to her. Again Ronnie picked her up and offered her the bottle. Again she refused it, twisting and turning, struggling out of the woman's arms. She pulled Ronnie's hair and bit her arms, not in fun this time, but violently. Ronnie put her down again, although this time not as gently as before. Patty Cake sat where Ronnie had put her and looked up into the woman's face. She seemed confused, reaching out to hold onto Ronnie's leg, but looking wistfully toward the door. She hesitated, then left Ronnie to go to the door again. She bent down and fingered the bottom, running her finger along the crack where the light shone through. Ronnie stayed where she was, sitting cross-legged, quiet and calm.

Uncertainly Patty turned to go back to her. She was undecided and then, halfway there, suddenly she sat down in the middle of the floor. After a moment or two of deliberation she opened her mouth and screamed. She waited, obviously listening for something, then deliberately screamed again. Again she waited.

At the first sound Caroline ran to the door, threw it open, and rushed in to her infant. As soon as the little gorilla saw her, she stopped her screaming, turned her back on her "mother," and strolled calmly over to Ronnie. There she sat, her hand resting on Ronnie's knee. At the window I burst out laughing. *This happened before*, I thought. *That smart little ape knows just what she's doing.*

But Patty would not take a bottle, and when Ronnie again tried to hold her on her lap, she began to struggle.

Patty yanked Ronnie's hair; long strands were torn out by the roots, clutched tightly in the gorilla's fist.

"Patty! That's enough!" Caroline took her up, and she sat peacefully on the woman's lap. She looked back and forth from one woman to another, unsure and suspicious, refusing the bottle again, hitting it out of Caroline's hand. At that Caroline decided that it was time for Patty's nap. As Caroline changed Patty's diaper, Ronnie straightened up the room. Caroline put her in the playpen, and while Patty lay clutching her radio to her chest, she leaned over her, cooing. Gently she covered her.

The three of us tiptoed out of the room. Just before we closed the door behind us, Patty opened one eye to see us go.

The next day Patty had diarrhea. Her expression was a bit more timid than it had been the previous day, and she did not come over gaily to greet me. Rather, she edged toward me nonchalantly, as if she did not know I was there. She sat on my foot as if it were an inanimate spot on the floor. Ronnie had taken charge, and when she began to change the little ape's diaper in her matter-of-fact way, Patty began to struggle, and bite. But Ronnie was stronger than and as determined as Patty, and the diaper was soon changed. Freed, Patty scampered toward her ponies and began to play.

She attacked me from the rear, climbing up onto my back and reaching around me to clutch my spiral drawing pad.

"No, Patty! You can cut yourself!" The spiral had long since begun untwisting from the holes in the paper, and the wire stuck out from one end. I pushed her away. Angrily she came back to hit me, clicking just as she used to with Lulu. She reached up and grabbed a handful of hair. Caroline came to my rescue, pulled her off me, and deposited her in the playpen.

"Now you behave yourself." The words were stern, but the voice was gentle. Sitting in the playpen, Patty looked wistful, and Caroline relented. "OK, Patty, you can come out. But be good now." I had a feeling that this was the only "discipline" that Patty received. Patty happily headed for her toys. But soon she came toward me and again tried to steal my notebook. Again Caroline put her in her playpen. She lowered the "ceiling" to make sure that Patty was in to stay for a while; then she and Ronnie went downstairs to the laundry room. Eddie and I were alone with Patty.

Things were quiet. I sat on the floor, taking notes and trying to sketch; Eddie stood idly, his eyes moving around the room as he leaned on the counter. Finally, very quietly, he made his move. Patty had not as yet truly accepted him, and now that the room was uncluttered and peaceful, perhaps this was the time for the man and the gorilla to become better acquainted. He bent over the play-

259

pen. And she came to him. As he knelt down, she stood up to meet him. Her tongue appeared as she concentrated on his face, and she tapped on the inner lining of plexiglass. He moved around the playpen; she followed him, circling her playpen. He stood up and left the room. Her eyes followed him as he went. He came back, and she greeted him. He lifted the lid of the playpen, and she reached up to him. He picked her up.

They were playing together when Caroline and Ronnie returned. Patty glanced up as the door opened, and she greeted Ronnie first.

She did not notice when Caroline left for the day; she was busy chasing Eddie around the room.

Ronnie shifted her position; Patty was distracted by the movement. She leaped up, laughing, reaching out for Ronnie's face, touching her nose and her cheeks. Ronnie sat very still, her eyes closed, as the little ape explored her face. Patty put her fingers into Ronnie's mouth very much as she had in Lulu's. It was a quiet exploration of this person who was to become the most important factor in her life for the next days. Suddenly she clutched Ronnie's hair in her hand and pulled. Ronnie pulled the giggling infant away. She held her up and somersaulted her in her hands, turning her over, over again, upside down, then right side up. "I'm gonna beat your ass!" She buried her face in Patty's stomach and shook her in midair.

Heh, heh, heh, heh, heh! Patty's giggles pealed forth uncontrollably.

Patty already reacted to Ronnie's presence as a natural part of her life, quickly transferring her dependence to her new surrogate mother. She accepted the food Ronnie brought her and trailed dripping cherry juice across the floor. She climbed onto my lap, the juice permanent stains on my drawings. She reminded me very much of her father, who had smeared so many of my drawings with urine. I popped a cherry into my own mouth, and before I knew it, Patty was climbing onto me, her eyes wide and eager. She peered into my mouth to see what was there and to watch me chew.

Ronnie hauled her away.

Ronnie was not as gentle with her as Caroline had been, but she allowed Patty to rough-and-tumble as much as she pleased. Patty would grab for her long hair, twisting her hands into it. As Ronnie straightened up, Patty's feet left the ground and she dangled in midair, swinging and swaying as Ronnie moved her head slowly back and forth. *Ronnie,* I thought, *must have the strongest scalp in the world.*

Patty played around us as if she had known us all her life. Yet sometimes, in the midst of her play, she seemed to change. She looked vaguely disturbed, as if she were interrupted by some reminder that things were not normal. She was distracted, for instance, by some noises outside the room and suddenly became quiet. She began to whimper. We could hear Eddie's voice, on the other side of the wall, chatting with one of the keepers from the Bronx Zoo. Patty knew the voice of the other keeper well. When the door opened and the two men entered the nursery, she hurried toward them. The keeper patted her on the head, and she stood up against him, hugging his leg, begging to be picked up and held. But he didn't oblige her. She watched him intently, and when he and Eddie left the room, she ran after them. She reached the door just as it closed. She tried to open the door, and when she found that she could not, she sat down and cried.

It was a few moments before she turned away and returned to Ronnie, her source of comfort and care. Once again a worried expression clouded her face. She wrapped her arms around Ronnie's legs and looked back first at the door, then at Ronnie, who was preparing lunch.

Again Patty refused to eat. She let the bottle fall to the floor. This time Ronnie did not pick it up, but she let Patty play with it until the gorilla decided to climb into her lap. Patty fiddled with the bottle, brushing the nipple against her teeth. One hand rested on the woman's knee. She sucked at the bottle once or twice, then stubbornly pushed it away and looked at me. She leaned out of Ronnie's lap and brushed the floor with her fingertips. When Ronnie put the bottle on the floor next to her, Patty climbed out of her lap to go to it. But she kept one hand on her new surrogate mother's body.

When Ronnie stood up to leave the room for a moment, Patty left the bottle to run to her.

"Perhaps she'll respond to something familiar," Ronnie said. "Hand me the lab coat, will you?" She slipped into the coat Caro-

261

line had always worn, sat in Caroline's chair, then bent over to pick the infant up. But again Patty merely played with the nipple of the bottle, grunting a bit until her head tilted back, her mouth opened a sliver, and her eyes closed. It had been an exciting morning.

"Don't worry," I said to Ronnie. "She'll eat when she's hungry. She's upset and maybe testing you, that's all. It's only the second day she's seen you." Ronnie put Patty on the floor. "Just look at her!" I said. Patty was merrily pushing a little cardboard cart across the room.

Patty did seem to be testing her new companions in many ways. The more I watched her display fits of temper until she got what she wanted and the more I watched her sullenness suddenly turn to eager enjoyment, the more I thought that she had been a bit spoiled these last weeks. If her negative behavior were ignored, perhaps she would begin to realize that her attitude made no difference in our treatment of her. Then she would eat.

When she again refused the bottle, Ronnie put it away calmly. No more was said.

We took her out for a stroll that afternoon, and when we returned and Ronnie left the room, Patty, true to form, began to scream. Ours was a deliberate experiment. For we were anxious to watch Patty's reactions to this very important part of her behavior. Patty, screeching, ran to the door. When it didn't open for her, she stopped her screaming and looked at me. But I went on with my business, taking notes, not reacting to her actions. She stayed by the door and once more threw back her head, opened her mouth, and let out a piercing cry. Again she turned and looked at me for a reaction. I showed none. I gave no indication that I had even heard her. She sat down. She seemed perplexed. This had never happened to her before. What would she do now? She looked as if she were thinking over the problem. Just as she decided to try it one more time and was about to utter another loud and piercing scream, the door opened and Eddie appeared. Ignorant of our experiment, he immediately picked her up and held her up to the window so that she could see Ronnie there.

Patty became furious. She struggled and bit and barked in his arms. Poor Eddie stood there, stunned, holding the angry gorilla. Ronnie, with a shrug of her shoulders, came back into the room.

Patty drank her next bottle. She held onto Ronnie, and before we knew it, she was done. I could not help thinking how different her life would soon be.

The following morning my husband accompanied me to the Bronx. Patty came scurrying cheerfully toward us. She had never met my husband, except as another face in the crowd outside her cages on Sundays. Yet she went to him immediately. He sat on the floor, and she climbed onto his legs to look into his face, then turned over on his lap and reached out for him with her teeth. She let him tickle her for a moment before she scrambled away and toward me. My greeting, too, was filled with her stinging little bites of affection. And she climbed up to grab the scarf that now protected my head from her fingers. She pulled it down over my eyes and tangled her hands in my hair until someone came to my rescue. I would have to find something else.

The day went quickly, full of fun and games with Patty. She took every bottle and ate all her food. She did not seem to miss Caroline at all. She greeted everyone who came in and giggled with pleasure as she sprawled on the floor. Fitz stopped by, as usual, to watch her, and suddenly he looked up and said, in his quiet and serious tone, "We'll take her home tomorrow." It was hard to believe that only three days had passed since the announcement that she was coming home at all. Ronnie leaned down and picked Patty up. She swung her high above her head. "Whooppeeee!"

CHAPTER 37
Bringing Patty Home
June 11, 1973

As I hurried down the path at the Bronx Zoo early that lovely morning, I hardly noticed the car coming toward me. But when it stopped a few feet away, a window rolled down, and an anonymous arm appeared waving at me, I had to look more closely.

There they were!

Fitz was smiling. For once a conspicuous gleam of excitement had slipped through his reserve. At the wheel Eddie was beaming. "Look what I got!"

I peered into the backseat of the car and saw Patty scrambling over Ronnie's lap and reachng for the cigarette in her hand. Ronnie took a puff; Patty reached up to touch the curl of smoke. As Ronnie exhaled, Patty stretched up to stick her nose in the midst of the stream. Then she saw me. She struggled eagerly over Ronnie's lap, dragging her plastic ponies with her, and climbed to the window to reach out to me. Her fingers tangled in my hair.

"Hey, don't fall out, pumpkin!" I laughed as I turned to Ronnie. "How is she? Do you think she minded leaving?"

"Minded? Look at her!" Ronnie had one arm around Patty's middle and was wrestling to contain the powerful wiggle. Fitz turned around to look at us.

"Come on," he said, "get in. You can admire her inside."

265

I ran around the car and swung open the door. Patty followed me, running over Ronnie and across the seat. I pushed aside her welcoming hands and slid in beside her.

"OK. Let's go!"

We left the Bronx Zoo behind.

As we drove through the city, I had a chance to look at her closely. The people at the Bronx Zoo had taken wonderful care of her. She looked beautiful and strong; her muscles moved flowingly under her black coat. Even as she made so simple a motion as reaching out to touch Fitz's face, her shoulder blades rippled under her glistening hair. Her eyes were shining and richly brown. She was sweet, playful, and curious.

Ronnie rolled down her window and air wafted through the stifling car. Immediately Patty leaned toward the breeze and stretched her arm out the window. She spread her fingers, letting the air rush over her hand, carrying it gently toward her face as if she could bring the wind in to touch her cheek. Her eyes were closed in enjoyment. As she played with the wind, we watched, smiling at her baby movements.

The morning rush-hour traffic was growing steadily worse as we neared Manhattan. As the congestion swelled, more and more people began to notice us. The man in the car next to us gazed sleepily out the window. His glance fell on us and then moved idly on, unseeing. Suddenly his head swung back violently. He stared in disbelief and idled his car closer to ours; he leaned out of his window, staring incredulously at Patty. She reached out gently toward him.

"I sure hope that guy wasn't out drinking all night." Fitz grinned.

On the highway we came to a complete standstill. Cars were backed up for miles. We sat in the sweltering heat with hundreds of other commuters. Every once in a while another car would creep toward us and pull alongside. The driver would dully glance over only to see a little gorilla in the back seat. The reaction was always the same: the unbelieving stare, then the shock of recognition. Inevitably the driver would pass his hand over his eyes and look again. Two gorilla eyes stared unwaveringly back at him. He would gape and smile until the traffic moved and another car would take his place.

We wondered if they knew who the little animal was. This morning's move had not been publicized. Fitz wanted Patty to come home with as little fanfare as possible to begin a quiet readjustment. The next days would bring many transitions for her to handle without the added problem of flashbulbs, loud voices, and probing cameras.

It was a long, hot, and frustrating ride back to Central Park, and Patty became restless. But we finally turned into the familiar cobblestone driveway in the park, and in a few more moments we were home.

As the door to Fitz's office closed behind us, Ronnie put Patty down. Within seconds she was scurrying around the floor, exploring. She found a pair of Fitz's shoes, almost immediately stuck her nose into one of them, then stepped into it. As she tried to walk, dragging the outsized shoe across the floor, we luxuriated in the sight of the unself-conscious gorilla, home.

She discarded the shoe when she caught sight of a wastepaper basket. She came rushing over to the desk, stood up, and peered curiously into the basket, high on her tiptoes. One hand went down into it and emerged clutching a handful of paper. It was only when the basket overturned and Fitz removed it to the top of his desk that Patty seemed to lose interest in it. She clambered over the rungs of my chair as if they were a jungle gym. When Ronnie left the room for a few minutes, Patty stopped to watch her disappear behind the door. She turned to Fitz and in a moment was secure on his lap. Immediately she leaned over to the wastepaper basket. He removed it to the floor, so she turned to his desk drawer, pulling it open and rummaging inside. She grasped one or two pencils and waved them in the air, dangerously close to Fitz's eyes. When he managed to wrestle them from her, she found herself once more deposited on the floor. The wastepaper basket was again transferred to the top of the desk. Nothing deterred her. If she couldn't play with one thing, there was always something else.

Fitz left the room, and she came to me, putting her hand on my arm and looking into my eyes. I felt grateful for her trust. "He'll be back in a minute," I reassured her. "It's OK."

The moment Fitz reappeared, she went to him, putting her arm around his leg and clinging.

"Hiya, kid." He patted her head.

267

"She missed you," I said.

"She did?" He looked surprised and turned back to watch her for a few moments. "You know, I never thought I'd see her here again," he said. "You don't know how happy I am that she's back."

When Ronnie opened the door, Patty ran to her. Then, secure with all of us present, she sat herself in the middle of the floor to play. Fitz seemed restless now. He put his hands in his pockets, then picked up a pencil, twirling it.

"Do you want to sit down?" I started to get out of the chair.

"Unh-unh." He grinned. "I'm too excited to sit down." He reached down to Patty, who was pulling on his trousers leg. "What do you want?" She looked up at him and lifted her arms.

"OK, come on." She grasped his hands, and he swung her high above him. Her mouth opened, and she giggled. Then he lowered her to his chest and leaned back against the desk, talking to her in a teasing, conversational tone. "What do ya think you're doing? Huh? What do you want? Ya want a little tickle?" She giggled. She reached out to grab his hair. "What do you think you're doing?" The words came out affectionately and easily. I had never before seen Fitz enjoy himself like this, all his hesitancy and reserve discarded. Patty pulled his hair and tugged at his sunglasses until they fell lopsided alongside his nose and hung precariously from one ear. He laughed as if for these moments he were totally released from inner restraints.

Patty ran to a suitcase and plopped herself down on top of it, pounding it like a drum.

"You like it? It's yours," Fitz said. Patty scurried over to his shoes and thrust her hand deep into the toe. She pulled it across the floor.

"You like the shoes? You got 'em." She dropped them and scurried off to look at something else.

When Patty's new nursery was ready, Ronnie carried her in. Everything was set up and waiting for her: the playpen which Dick Berg had bought for her, her old toys from the Bronx Zoo and some new ones. When she saw her little yellow truck, she headed right for it. She climbed into it and began to roll it along, pushing her hand along the floor to move it.

"Do you think she's ready for a bicycle yet?" Fitz murmured.

"She really has nothing to swing on," someone remarked, and Fitz disappeared. In a few moments he came back and hung his belt from the bars at the bottom of the window. Patty watched him curiously. When he was finished, she sniffed at the leather, then climbed the belt immediately, swaying and swinging to and fro.

Ronnie and I settled in, and visitors began to arrive: a beaming Luis, who winked at me happily, an excited Richie Regano. They were pleased because they felt not only that Patty belonged there, but that they had been exonerated from the implications that they had not properly cared for her. Other keepers arrived, then people from the commissioner's office, and finally Commissioner Clurman himself. Patty looked up from her toys and scrambled toward him. She put her arms around his leg. I reached over to help him extricate himself from her grasp. He looked at me strangely as I grabbed hold of her and drew her away. "I just wanted to make sure she was all right," he said, one pants leg a little more wrinkled than the other. He gave me one last questioning glance before he disappeared, a glance that I could not understand—until a few minutes later, when I put my hand up to my head. To protect my hair from the gorilla's fingers, I had put on my pink bathing cap, the one with the flowers on it. I had forgotten all about it.

Patty was very excited and overly active. She fought when Ronnie changed her diaper, and she refused her bottle, wanting to wiggle off Ronnie's lap and down onto the floor again. She refused to stay in the playpen, but kept climbing over the top. But as soon as Fitz had secured the mesh lid to the playpen, Ronnie deposited her inside.

"Stay there, P.C.," she said firmly. "Lie down. Go to sleep." Finally, Patty stopped struggling. Soon she lay down, and within seconds she was asleep. Then Ronnie and Fitz left to do some paperwork and have some lunch.

When Patty woke, she and I were alone in the room. "Hi, Patty." She stood up and looked at me through the green and orange bars. Half her diaper flopped off and hung lopsided. I changed her, not a very good job my first time diapering a wiggling gorilla. But when the diaper was finally on, she reached up to me, whimpering, begging to be taken out of the crib. I picked her up, and she clung tightly to me. I heard a little, hardly audible whimper.

"What's the matter, pumpkin?"

269

It was the first time in nine weeks that she had awakened away from her nursery in the Bronx, the first time that she had not seen Caroline Atkinson's gentle face looking down at her. She clung, her grip tightening on my arm. I talked to her and showed her where she was. I showed her all the new toys and her old familiar, comforting ones. She nuzzled against me, and I stroked her back. Her fingers pinched as she clung. Little by little she relaxed, and soon we were sitting on the floor, playing with her busy box. By the time Ronnie returned Patty greeted her cheerfully, scampering to the bars as soon as she heard the door open. Then, as Ronnie gave her a bottle of Enfamil, I left to pick up a folding cot.

Someone was to sleep in the room with Patty for as long as she was here, so I borrowed a cot from my parents, who lived near the zoo.

The cot was too cumbersome to get on the bus, so with great difficulty I wheeled it through the elegant East Side streets to the zoo.

Finally, I wheeled it down the cobblestone walk and triumphantly into the office. As I opened the door and pushed the cot through ahead of me, Fitz looked at me with a twinkle in his eye.

"Next you'll be bringing us chicken soup," he said.

CHAPTER 38
Patty and Darwin
June 12–15, 1973

The following morning Ronnie arrived with a black carrying case. From its interior came the shrill twitterings of a familiar voice. A flash of a tail swirled by the bars of the case. And Darwin's face appeared, excited and nervous.

Soon the monkey streaked out into the room. She bounded to her surrogate mother's shoulder, squeaking deafeningly into her ear.

Patty Cake stopped in her tracks to stare.

Darwin's appearance at the nursery was not coincidental. She was to be a part of Patty's transition back to animal life.

Although Patty had been in the presence of Mopey and Hodari, she had seldom interacted with them. It was important that she become used to the casual interaction of animals. For soon, if everything went well, she would again become part of familial intimacy.

For more than an hour Patty roamed and Darwin darted, and each secretly watched the other from the corners of her eyes. Finally, unable to resist the whirling antics and flashing tail of the monkey, Patty watched openly and tentatively reached out to touch the quicksilver body. But by the time Patty's hand reached the spot where Darwin had been, the monkey was already somewhere else. Amazement halted Patty in her tracks, and then she

271

was off! She seemed fascinated by Darwin, and suddenly nothing else seemed to exist for her; the toys were left alone, and she didn't even want her bottle.

How different they were, the Old World ape and the New World monkey! Darwin could fly! As weightless as a bird, she bounded out of reach, her thin body stretching, twisting, bending, landing, then spinning out again while a shower of squeals rained into the room. Patty seemed earthbound. She paused, confused by a sudden change of direction, then scrambled after Darwin again. She tripped over herself in her ungainly gorilla babyhood, then toppled nose first to the floor. She shuffled along, giggling, her head tilted back, her eyes fastened on the ever-spinning monkey. Suddenly Darwin sprang toward her, touched her lightly with her foot, then was off again, all in less than a second.

Even as Patty's gentle fingers closed over Darwin's tail, Darwin escaped in an effortless bound—until Darwin was not quick enough when the hand descended. Darwin, caught by a leg, spun toward Patty, who hugged her for a moment before Darwin leaped away.

Patty giggled and slid across the floor after her quarry.

The room and everything in it became part of their play. Even I found myself a part of it. Darwin sprang to my shoulder and caught her matchstick fingers in my hair, whistling, as she sucked on my earlobe. Patty circled us below, then stood up against my legs. Reaching up toward her new friend, she pulled on my clothes, trying to climb. She wrapped her arms around my legs and started up. Suddenly the monkey sprang away, strands of my hair still clutched in her tiny fingers. And just as suddenly my legs were freed.

Darwin pounced on Patty's string of plastic ponies and dragged them after her.They caught in the mesh of the playpen lid and rattled and clattered against the metal. Just as Patty's hand reached out to grab them away, they were free and went flying out behind Darwin as she soared away. Patty turned and looked after them, then determinedly plodded along after them. Finally, Ronnie managed to rescue them. Patty snatched them out of her hand with a possessive swipe and went shuffling off, dragging them behind her across the floor in a defiant, resolute march. Darwin vaulted to-

272

ward the gorilla. This time Patty lunged toward her, grabbed her, and bit.

At Darwin's squeal we separated ape and monkey. In a few minutes all was amazingly quiet as Patty in her playpen and Darwin in her cage nodded off and went to sleep.

Even in sleep they were different: Patty stretched out full length, one foot sticking out between the bars; Darwin lay curled up into a little ball, her tail a final curve that wound around her body, as compact and as unnoticeable as possible.

When Patty woke and Ronnie lifted her from her crib, she trotted straight to Darwin's cage. Darwin, seeing her there, bounded to the bars and latched on, chattering at the little ape, whose tongue peeped out at her in concentration. As soon as Ronnie opened the door, Darwin leaped out into Patty's arms.

They played wonderful games of tag, Patty giggling with glee. She lay on the floor, trying to keep track of her new friend. Darwin used Patty as a stepping-stone. She sprang up onto a table or chair, and Patty did her best to catch her in midair. When she did, Darwin's arms and legs wiggled and waved as Patty enclosed her in a gentle hug. For the first time in a long time I heard Patty laugh without any direct physical stimulation. Just watching Darwin as she leaped and flitted and soared around her, simply enjoying their interaction and their closeness, brought it on.

As they began to tire, the illusion of a multitude of monkeys dwindled and dissolved into one sleepy capuchin. They moved more and more slowly until finally neither one of them could stay awake. Darwin dozed off quietly in Ed's arms, and Patty, who was padding sleepily on the floor, simply came to a halt, put her head on her arms, and fell asleep where she lay. Ronnie carried her off to bed; Patty never stirred.

When Darwin woke, she twittered and squeaked, impatient to be released from her cage. Then she went looking for her friend. She vaulted to the playpen and peered in through the bars, hopping impatiently from one slat to another. Patty slept on. Darwin wanted to play. She slipped between two bars and poked the gorilla's arm. Nothing. She leaned over and nudged Patty's shoulder, squeaking shrilly into her playmate's ear. Patty Cake stirred, opened one eye, then reached out toward Darwin. Quietly they began to play. The slim little red monkey slipped in and out of the playpen like mercury; when they met, Patty hugged her, giggling. Once, when she bent over the monkey, she reminded me of Kongo when he had bent over his daughter and kissed her. She looked so much like him.

Patty's acceptance of Darwin and her display of affectionate enjoyment represented far more than the momentary pleasure that these animals were so obviously experiencing. It was bringing Patty closer and closer to a reunion with Lulu and Kongo. And with each responsive gesture we grew more confident about what we would soon attempt.

CHAPTER 39
In the Lion House
June 12–14, 1973

The very morning after Patty's return to Central Park she was taken to the Lion House.

Lulu and Kongo were put outside in the corner cage that morning. There was no way that they could know she was there, no reason for any premature excitement or frustration for them. But their scent was left behind purposely; a towel that had been rubbed across the floor was deposited in the number three cage, where Patty would have her first glimpse of her old life.

What would her memory hold? Would she know the smells? Would she remember the bars? Would it mean anything to her at all? Would she, could she, readjust to a life without colorful toys and sheet-covered mattresses? Without people directing her, feeding her, playing with her? Other than her actual responses to her mother and father and their acceptance of her, these were our most prominent questions.

Ronnie carried her in and sat with her as she made her first exploratory movements. She did not seem shy or fearful; rather, curious and interested, she went directly to the center pole and began to climb. Then she scampered over to the bars.

When Eddie, standing outside the bars, called her and rattled his keys, she went to him and played with his fingers between the bars.

275

Then she turned away to explore. After twenty minutes of more and more casual play, Ronnie and Eddie picked her up and took her back to the nursery.

She had shown little of the anxiety which we had anticipated. Had she remembered her parents? She showed no sign. She had not gone looking for them. Had she remembered the day of trauma, the day of the accident? She showed no fear of the bars where it had occurred. It had been a satisfying morning.

The following morning we again carried her into the Lion House. We took her plastic ponies and her towel with us. They would be familiar objects in what might seem a strange place to her.

Again Ronnie carried her inside. Again Patty went to the center pole, her ponies clattering behind her. Soon she headed for Eddie, who was talking to her from the keeper's walk, and then, distracted by something under her feet, she forgot him and bent down to examine the cement. She ran to Ronnie, who sat cross-legged in the middle of the cage, and then left her to go again to the pole. She went back and forth to her surrogate mother, interrupting her play for the cuddling reassurances of her presence. When Luis passed by the cage on his way down the aisle, he stopped to say hello to her. It had been a long time since she had seen him in the Lion House, but she went paddling over to him and reached out to him. He took her hand and shook it, smiling. "Hello, my little one." She took an apple from his outstretched hand.

Ronnie took Patty from one cage to another. Patty seemed completely at home. She seemed to feel safe and secure until a sudden strange noise sent her scurrying back to Ronnie. But, I thought, she most likely would have run to Lulu had she never been away. Her fear forgotten in a second, she went toward Fitz, who stood quietly outside the cage, tapping his hand on the floor to call her over to him.

It was very important that she be trained to come to the front of the cages, for there she would be receiving supplementary feedings of formula, food, and juices. And training her seemed easy, now that she was used to being fed by humans. She started toward him but became involved in examining a spot of paint on the floor. Excited, she scurried back to Ronnie. Again she started out toward the bars and the bottle. She tapped her fingers on the bottle, then

276

scurried back to Ronnie. During the third journey toward the bottle she became distracted by a puddle of water and then by a rather large water bug. She stopped and stared, intrigued by the creature that moved directly past her nose. She turned and followed it through the puddle, then reached out to play with it. She pushed it farther and farther toward the bars, toying with the bug until she accidentally pushed it too far. It toppled into the trough. As it was carried away, she sat at the bars, her hands gripping the poles, watching it go.

Then she turned to Richie, who had joined us there.

While they played, Ronnie rose and, careful not to catch Patty's attention, moved into the number two cage. In her fascination with the water bug Patty had left her string of ponies in the center of the cage. She must have remembered them, for she turned and reached out. Suddenly she realized that Ronnie was gone. She quickly looked around the cage and began to whimper. She stumbled toward the door to the next cage, scrambled in to Ronnie, and for a few minutes she clung to her insistently, whimpering if Ronnie showed signs of putting her off.

"Hey, P.C. Whatsa matter? It's OK. I'm here." Ronnie's hearty voice soon satisfied Patty. Little by little she began to play. Soon she had become comfortable enough to leave Ronnie and venture out on her own.

Now Fitz beckoned to Ronnie, and very slowly she moved toward the front of the cage, ducking down and hopping to the ground. The cage door closed behind her. Patty was alone.

Patty ran to the bars. She reached imploringly out toward Ronnie, and Ronnie touched her hand. Suddenly Patty looked so small and lonely as she sat, clutching the bars, looking soberly into Ronnie's face. She climbed a little to look directly into Ronnie's eyes, solemn and still; all her energies centered now within herself (she did not reach out as she had before). She gazed with big, round eyes toward Fitz. "What is it, P.C.?" he said, and she crossed over to him, then back to Ronnie again. As Ronnie deliberately put her foot up onto the cage level and one hand on the bars, Patty grabbed her finger and held it fervently to her chest. She jumped at a strange sound and stumbled to the door. She tried to open it, to pry it away from the bars, pulling with all her might. But it didn't

277

budge, and she again squatted in front of Ronnie and reached out to touch her face. She was not interested in the bottle of Enfamil that Ronnie offered her but wanted only to hold onto the woman on whom she relied for so much. She was absolutely quiet and motionless. The space in the cage seemed to engulf her. The bars loomed up higher and higher. And she sat taking up so little space, clutching a bar with one hand and Ronnie's finger with the other. Her towel trailed out on the cement behind her, forgotten. Her ponies lay discarded. Her eyes gazed beseechingly on one thing, Ronnie. They seemed to become more and more anxious as the seconds passed.

At Fitz's orders, Ronnie opened the cage door. Patty scrambled toward her, almost tripping. Ronnie lifted her out of the cage. Patty clung to her, snuggling against the woman as closely as she could, whimpering while Ronnie hugged her and soothed her. Patty's apprehensive eyes peered into mine over Ronnie's shoulder as she was again carried safe and sound down the corridor of the Lion House, out into the sunshine and toward the nursery and Darwin. We had been in the Lion House for nearly an hour.

At the nursery Patty turned happily to her toys and charged after Darwin as the monkey spun around her, excited to see her friend. Within a half hour Patty had consumed eight ounces of Enfamil with cereal and applesauce and scrambled to the floor to play in carefree joy. None of the anxiety she had displayed at the Lion House remained. She trotted over to the strap that Fitz had strung up for her to swing on and was trying to untie the string that held it when Herschel Post, a member of Commissioner Clurman's office, came in to see her. His eyes were filled with sympathy and worry. "I heard that she has brain damage," he whispered to me.

"What? Where'd you hear a thing like that? Just look at her!" At that moment she had succeeded in untying the string. The strap flopped down over her back. She grabbed its end and began to wind it around her leg. Mr. Post looked at me and laughed in relief. She looked up at the sound and padded over to him in greeting.

"I'll never believe anything like that again," he said.

That night there was a thunderstorm. Lightning flashed and thunder crackled around us. The rain poured, splattering the window in the nursery with a continuing clatter. We had been told that Patty was terrified of storms, and we waited anxiously for signs of her fear. But there were none. She merely climbed up on the strap that hung under the window to look curiously outside. That was all.

The next day Ronnie entered the cage with the clinging infant again. When she opened her arms, Patty climbed out of her lap and strolled happily to the bars. She showed no reaction to her experience of the day before—no fear and no hesitation. She wandered around, dragging her ponies after her, poking her nose into the corners, and putting her hand into the water that flowed down the trough. The noise of the men going about their usual business seemed to make no impression on her. It was possible, I thought, that her continual glances and trips toward Ronnie might have been not a result of insecurity alone, but simply the method by which an infant of growing independence keeps track of her mother in the wild. After all, Patty was not yet a year old. Lulu would still be keeping her close to home, anxious to keep her eye on her roving child as if, in the wild, she were about to move on to another feeding place. Patty's need for Ronnie's presence might have been emphasized by her loss of her real mother and then of her first surrogate mother, but perhaps the basic training which Lulu had given her had been strong enough to carry over these weeks of separation. Or perhaps Patty's need for a mother was so basic a drive that it was deeply embedded in her.

On a cue from Fitz Eddie climbed into the cage, played with Patty for a minute or so, then went into the adjoining cell. He shook his keys. At the familiar sound Patty stopped swinging from the crossbar and hung there, listening. She tilted her head toward the number two cage. She saw nothing there. When Eddie rattled the keys again, Patty lowered herself to the floor. She strolled toward the door, then stopped. Again he clinked the keys, and Patty stood uncertainly, listening to them, her head inclined toward the noise. She seemed to be thinking. Suddenly she made up her mind, and with no further hesitation she trotted in. It was the first time that she voluntarily left the cage in which her "mother" sat. In a moment or two Patty reappeared in the doorway, stopped for a moment, then continued on to Ronnie. She walked into the keeper's lap, touched her collar, then turned and skipped happily back to

Eddie. She scampered back and forth, perhaps making certain that each of her friends was where she had left him, until finally, tiring of the game, she was diverted to the bars, where she again bent down to dip her fingers in the water.

Then her eye was caught by Princess, the tigress whose cage was across the aisle. The undulating movement of the cat seemed to have a hypnotic effect on the little gorilla, for as Princess paced back and forth along the bars of her cage opposite Patty, the gorilla stared, fascinated. She climbed the bars a foot or so without taking her eyes from the cat, then climbed down again, almost losing her footing. She reminded me of her mother, her eyes alive with intense concentration on the movement across the way. Once or twice her eyes flashed back to Ronnie, who remained sitting quietly, watching her charge. But her glance was drawn back to Princess as if the cat were a magnet. Eddie crawled up behind Patty and called to her, tempting her to play, but Patty showed no reaction. She squatted at the bars, one hand dangling limply in the trough, her eyes seemingly aware of nothing but the tigress. She sat motionlessly for a long time, mesmerized. Then Ronnie called to her. "Hey! Whatcha doin,' P.C.?"

Somehow the spell was broken. Suddenly Patty became aware of her hand that still dipped into the water, and she turned it toward the flow in order to catch a bit of garbage that was floating by. She was normal again.

She seemed far less dependent on Ronnie. Fitz decided that it was again time for Ronnie to slip out of the cage. Again Eddie entered to distract Patty while Ronnie sneaked away. He picked up the little gorilla and swung her playfully above him. *Ho hooo. Ho hooo*, he sang out. She giggled. When he put her down, she cheerfully trotted after him and followed him into the next cage, trying to grab at his ankles.

"*Ho hooo* . . . come on, Patty, you can't catch me. . . ." She adored the game of catch and tag, and they disappeared into the cage, laughing.

That was Ronnie's cue, and at a nod from Fitz she left the cage. When Patty came to the doorway to check up on her, the cage was empty.

Patty forgot all about Eddie and their game. She ran into the cage, her head turning as she looked for her "mother." When she reached the center of the cage, she stood still, swinging her head as she searched the bars and the corners. Confused and disconcerted, she stood very quietly as if she did not know what to do.

"Hey, P.C., look what I got for you!" Ronnie said as she reappeared and showed Patty a bottle of formula. Patty rushed toward her. She ignored the bottle in Ronnie's hands and instead tried to reach her. She pressed her body against the bars, getting as close to Ronnie as she could. Her hand reached out for her. Ronnie slipped the bottle against Patty's open palm. She grasped it and timidly, obediently, squatted at the bars and drank, gazing soulfully at Ronnie.

Fitz and I looked at each other and nodded. We knew now that Patty would be able to receive her supplementary feedings. Whether she drank because she was obedient or whether the bottle was familiar and comforting to her or whether she was hungry we did not know. But she drank. Still, Ronnie was not the only person who was to feed her, and in another moment Eddie, too, crept out of the cage and joined us at the bars. He took a bottle from Ronnie and offered it to Patty. Subdued and quiet, she took another ounce or so from him.

As she began to realize that she was now completely alone in the cage, she stopped drinking. The bottle slipped out of her grasp onto the cage floor. She began to whimper. She stood at the bars, looking from one of us to the other, turning her head and shoulders to look back of her into the empty cage. Then she turned to us again. She crept toward Ronnie and crouched down to touch her shoe, which rested on the trough. She pressed her finger into the leather and looked up into Ronnie's face. Then she stood up and turned around, looking, searching. She sat down and touched the floor, a vague and timid gesture. She rose and moved her fingers over the diamond-shaped mesh at the edge of the cage in a distracted way, then wandered back to the center to sit in front of Ronnie. She let her hand dangle in the trough for a second or so before she once more reached for Ronnie's shoe. But this time Ronnie deliberately took her foot away, and Patty's hand slipped to the floor. Ronnie walked away. Patty looked up with hurt eyes, and my heart went out to her. All this was for the best. But she did not know that.

281

Her movements were quiet and hesitant. She stood up and wandered over to the mesh again, then back, climbed a foot or two, then came down. Glancing back, she saw her towel lying on the floor where she had left it. She rushed over to it and sat down, holding it tightly against her chest and putting one corner to her mouth. Did it alleviate any of the growing insecurity that she felt? She had developed the habit of thumbsucking during her stay at the Bronx Zoo. So she clutched and sucked on the towel as she had clutched and kissed and bitten Lulu, Caroline, Ronnie, Fitz, and myself. She would not let it go. Except once, when she saw Eddie move toward the door. Then she dropped it and rushed in that direction. But he only walked past the door and on. As he disappeared from her sight, she turned and scurried back to her towel. Now she carried it with her, dragging it along between her legs, as she came to the front of the cage. When Ronnie moved, she followed; then she seemed to have an idea. She went to the door and tried to pull it open. The towel was left behind; she tugged on the unyielding metal bars with all her strength. But when she realized that the door would not open, she took up the towel again and clasped it. First she merely held it. But as time passed and we simply stood outside her cage watching her, she surrounded herself with it, chewing on another corner. Her eyes were wide, timid, and wistful. She lay down and rolled over to look into the empty space behind her again, then sat up, and reached out for the bars. She began to climb and reached behind her for the towel, which she now would not leave at all. There seemed to be no purpose in her climbing; it was just a haphazard activity. Her body was doing something. The towel caught on a bar. Patty stopped to pull at it, and as she did, her attitude changed. She tensed. I could see an idea strike her. She slid down the bars, dropping the towel, and rushed to the door between the cages. She looked excited, and with tremendous energy she thrust herself toward the door and looked in. I knew that she expected to find someone there. This had been a long game of hide-and-seek. She was again playing. But the cage was empty.

Uncertainly she looked toward us again, and her demeanor changed. She slumped and then dragged herself sadly toward us again, disappointed. She went to her towel and lay on it, staring up at the center pole above her.

Ronnie laughed at something someone said, and Patty's head jerked up at the sound. She came to the bars just as Ronnie appeared at the edge of her cage. Once more she obediently squatted before her and accepted a bottle of juice. Then Ronnie slipped out of her sight again.

A door slammed, and the infant jumped.

Again she sank onto her towel.

When she caught sight of Ronnie and Fitz she began to whimper. She followed after Ronnie, yet when the woman stopped before the cage, Patty turned her back to her. She sat facing the wall, turning her head ever so slightly every once in a while to look at Ronnie from the corner of her eye. The towel was tightly embraced.

When Ronnie opened the cage door to let herself in, Patty sped toward her and grabbed her leg.

"Hey, wait a minute, P.C." Ronnie could hardly move. "Hold it." Patty was clinging as closely as she possibly could. Ronnie stumbled and sat down while eager arms encircled her and Patty scrambled onto her lap.

It was a few minutes before Patty began again to reach beyond Ronnie's body. And then she kept one foot or a hand on her person. If Ronnie shifted her position, Patty was back, climbing onto her lap, pulling at her clothes. But finally, the gorilla began to relax. She leaned comfortably against Ronnie's leg, and then, after two false starts, she became brave enough to leave her altogether. She went to the door and walked around it. Swinging it back and forth on its hinges. Then she ran back to Ronnie and threw herself at her. Soon they were playing. Patty was giggling and free, her fears forgotten. The next time she ran to Ronnie from the center pole which she had climbed, Ronnie picked her up and carried her out.

For a full half hour Patty had been alone in the cage.

Now it was back to the nursery and Darwin.

Later that day Dr. Nadler arrived from Georgia. While Patty and Darwin slept, nestled in each other's arms, we met in Fitz's office to discuss the reunion. We covered my notes and Ronnie's reactions to Patty's responses. We discussed the possible responses to the actual reunion, the logical progression of adjustments to be made, and Patty's natural trepidation. Too, we discussed methods of separating mother and child should there be difficulties at their meeting.

Dr. Nadler seemed confident and happy with Patty's reactions so far. He was amazed at her easy adjustment to Ronnie and her lack of concern over Caroline's sudden disappearance from her life. He was impressed by her immediate acceptance of the new nursery, Darwin, the cage, Sharon Fama and her whirring cameras, and her new companions. As we led him in to see her, he was surprised by her lack of fear of a stranger. She came trotting toward him to investigate his smell. She showed no apprehension, sitting calmly between my feet, intrigued by my shoelaces. She showed no sign of missing Ronnie when the keeper left the room; indeed, she barely looked up to notice.

"I don't think it's necessary to wait any longer," he said. "In fact, the sooner we put her back with her mother, the better."

He went on in a maddeningly conversational tone. "Her reactions seem to be normal. I don't think they'll change or improve at all. It seems senseless to let her get used to any more . . . uh . . . I don't think that it is necessary to—" There was another professional pause. Pipe smoke swirled up into the air.

"What about tomorrow?"

CHAPTER 40
The Reunion
June 15, 1973

"Say good-bye to Darwin, P.C."

Unconcerned, Patty swung on the leather strap under the window. It swung to and fro for a moment before she dropped to the ground and headed for Darwin's cage, familiarly fingering the thin rods of the bars. The capuchin bounded toward her, squeaking and whistling excitedly from behind the bars. She was not to come out now. They had played earlier this morning for the last time. Patty ran her fingers between the bars, trying to reach her. She put her tongue between the bars as she had so often, her eyes wide and eager as she watched her nervous little friend reaching out for her. Darwin twittered ceaselessly.

We sensed some movement behind us and turned. Fitz was silently watching.

It was time to go.

Ronnie and I looked at each other. Oh, God, this was it!

As we walked down the hall, Darwin's excited chattering grew fainter and fainter. The door closed behind us, and it was gone.

The press was waiting; they surrounded us as we began our last walk toward the Lion House. We walked by the llama that had been born that night; delicate, curly-haired, deep chocolate brown and white, she was already following her mother on fragile, thin legs. We walked on past the sea lion pool. Fitz led the way. Ronnie followed, Patty, wrapped in her towel, held in her arms. Then I came, still nervously taking my notes. Parks Department officials joined us. Reporters strung out behind us, and cameramen jumped in front every once in a while, then lagged behind for another angle. Patty twisted in Ronnie's arms and reached out to them.

People who had been strolling around the zoo stopped to stare at us wonderingly. Then someone realized what was happening.

"Look, there's Patty Cake!"

"Patty Cake, Patty Cake!" A woman picked up her child to get a better view. A small crowd began to gather alongside our parade. We never lost our pace but kept going.

Suddenly the circle of reporters and photographers broke, racing into the Lion House ahead of us. We found ourselves suddenly standing alone.

"Well, girls"—Fitz looked down at his feet for a moment—"I guess this is it." He was as reserved and as restrained as ever. "Thanks for everything," he said. Then he led us inside.

We were stunned! No one had told us that there would be so many people! It looked as if all the New York press were there. Cameras, lights, microphones and wires, cables and boxes lay strewn on the floor. Skip Garrett was running from one person to another, relaying information. The doctors from Flower and Fifth Avenue Hospitals and, of course, Ed Garner, stood discussing the case.

Suddenly, they were ready for Patty. Fitz waved us forward into the midst of the crowd. As we walked toward them, I could not resist touching Patty for one last time, and my hand lingered on her hair. Her shining eyes peered into mine inquisitively. And then she was gone. Ronnie climbed over the railing into the keeper's walk.

Quickly I wove my way through the crowds to my place in front of the cages. All the gorillas' cages were empty. Lulu and Kongo were in separate cages outside. Even Caroline and Joanne had been put outdoors just in case their cage, too, was needed for the reunion. As I settled down, Eddie opened the door to the number three cage, as he had for the last three days and, as usual, Ronnie ducked low and carried Patty in. Again she went to the center of the cage and sat down, cross-legged. She put down Patty's towel and then, for the last time, removed the diaper that Patty wore. Patty would not need the diaper again.

Patty squirmed in Ronnie's grasp, impatient to play, and Ronnie let her go. The gorilla headed straight for the center pole, ignoring the large semicircle of apprehensive, whispering people before her. She toddled over to the opened door and splashed happily in a puddle that she found en route. Without the Pampers her stomach looked rounder than ever, and as she bent over to look at her reflexion in the water, I thought, *She looks just like a precise miniature of Kongo.*

Eddie tried to tempt her to the front of the cage with a bottle, but Patty ignored him, enjoying her pool of water. Then, distracted by the open cage door nearby, she scurried over to it, stood against it, and swung it slowly back and forth on its hinges. Ronnie sat very still, biting her lip in her anxiety. She welcomed Patty when she came to her and gave her a loving hug. Then she let her go.

At the bars Eddie shook his keys at Patty Cake as she hung from the crossbar. She stopped wiggling, then scrambled down the pole and went to him, reaching out. At Fitz's signal Ronnie took the opportunity to leave the cage. It was ten thirty.

From my diary:

> She is alone in the cage now, and all the whispering activity is outside. She looks so quiet and so vulnerable on the floor, sitting on her towel, fingering it the way she did yesterday. She knows that Ronnie is gone, for she looked up suddenly and saw that she was not there. And she did not know what to do. So, like yesterday, she turned to her towel and became quiet as her fingers ran over it. Every once in a while she tilts her head toward the cage door, closed and locked now. But she makes no move toward it. Still innocent and so, so small.
>
> It was so hushed now, as the time came closer. Eddie stood on the trough by the number two cage, his hands on the ring, ready to pull the chain that would open the door at Dr. Nadler's signal. Fitz knelt in front of Patty on the left side of the cage. I could see his mouth moving as he spoke to her. She looked into his eyes as she quietly held onto her towel.
>
> She does not know.

At ten thirty-five Eddie opened the door to the outside and let Lulu in.

She had been hovering at the door. She and Kongo had been separated that morning and that in itself was very unusual. She knew that something strange was happening. When the door was opened, she sprinted in purposefully and eagerly. She went directly to the mesh wall between the cages and looked in.

And screamed!

Her first piercing shriek cut through the silence of the house.

"She sees her! She knows her!"

Lulu darted back and forth along the wall, her eyes riveted on her daughter. In her excitement she let out high, shrill shrieks that went on and on. She sprinted back and forth, unable to keep still.

At the first shriek Patty's head had jerked toward the sound. Her head tilted toward the quick shadow that moved in such a frenzy of excitement. Then Lulu bounded toward the door separating them. As she stood shuffling, waiting impatiently for the door to be opened, Patty stumblingly followed the shadow toward the back of the cage.

Eddie began to ring his keys frantically. It was imperative that Patty be at the front of the cage. The men had to be able to reach her if anything went wrong. At the familiar tinkle Patty paused. She looked toward the sound, distracted, hesitated, and came toward Eddie. We sighed in relief.

On the other side of the screen Lulu screamed again, and as she lunged down the length of the wall beside her daughter, Patty became aware of the rush of the movement. She looked up and froze at the sight of the frantic, hooting shadow. She, too, began to scream. She fell to the ground and half crouched in terror. The sounds of terror sent shock through the people who watched.

Dr. Nadler shook his head. "We can't wait any longer," he said. "She's too excited."

He signaled. "Now!"

Eddie pulled the chain.

Lulu bounded through the door. At the sight of the sudden violent movement behind her, Patty crumpled to the ground. Her legs collapsed under her. Her arms folded, clutching her chest; her face turned to the hard floor; she cowered, terrified. At the sight Lulu

288

drew back. Her surge forward was suspended, and she stood motionlessly, watching. She took one deliberate step forward, stopped, then moved again, one step at a time. With unbelievable restraint she came to her child. Tenderly she squatted down beside her trembling baby. Slowly and tentatively she moved her hand above the infant's back. The baby screamed! Lulu withdrew her hand. She stared down at Patty for a moment, then stood up and walked away. She came to Patty's towel, picked it up, smelled it, then put it down. She wandered over to the side of the cage, then turned and walked back to her daughter, bent over her, and put her lips to the baby's head.

Patty screamed! Lulu sprang back.

Again she gradually moved her hand over the baby. But she did not touch her. Again Patty screamed in fear. She defecated; her pale, loose stool glided over the cement. Lulu gently put her hand onto her baby's back as if to comfort her.* But Patty screamed again and clutched the bar next to her, squeezing herself against the bars, as close as possible to the outside, as far as possible from her mother. If there was any way for her to get out, she would have squeezed her body between the bars. A little moan escaped from her. Lulu gazed down at her child for a second or so, then turned, walked toward the door that led to Caroline and Joanne's cage and went through. She climbed onto the ledge to look out the window. On the other side Caroline and Joanne shifted about, trying to look in on their cage.

Timidly Patty raised her head and loosened her grip on the bar. For the first time she sat up and turned her head to look after her mother. But at the sight of Lulu returning, she bent once more to the floor, clutching her body. Feces slid over the cement. Another pathetic moan came from her.

Lulu put her hand softly onto Patty's back, as if to quiet her. Patty seemed to press her body into the cement. Once more Lulu left her, this time to go to the door that led to Kongo's cage. She seemed to be looking for him, for she swung up to the window to look outside for a moment. But she could not see him and soon came down again. As she came back toward Patty, she came

*Was this the usual and identifiable sign of reassurance?

289

across the towel again. For the second time she picked it up and smelled it. She rubbed it across her face and examined it, holding it at arm's length. Then she threw it toward the edge of the cage, where it lay discarded, spilling out between the bars and into the trough. She turned and deliberately marched to her daughter, leaned over her, and kissed her.

Still, she would not pick her up. She bent down to her, moved away, came back, and left again. The trembling baby shifted her position slightly and followed her mother with fearful eyes.

Lulu obviously yearned to touch her, to take her. Yet she desisted. Tentatively she reached out again and again to touch Patty or to take hold of one of her rigid arms, then withdrew as Patty's body jerked in involuntary recoil. Patty scuttled backward on her stomach as Lulu bent over her. She tried to sink deeper into the floor. Lulu's hands often went out to her baby, but each time, when she saw Patty's fear, she withdrew her hand. She waited for her baby to accept her. She watched Patty huddle against the cement, whimpering and trembling, her hands covering her head. After all these weeks of grief and suffering, the mother sat patiently and waited. It was a wondrous compassion, a compassion we had not, could not have, predicted. Her baby's needs were so much more important than her own.

Little by little the miracle happened. Five minutes after they had first seen each other Lulu touched Patty's arm and Patty did not scream. As if she knew that she handled a precious, frail, and frightened being, Lulu gently lifted Patty up and slipped her onto her own body. The baby did not cry out.

Lulu held her child enclosed in her arms for the first time in weeks. And looked softly down on the little, still, rigid body that she pressed into her own long hair.

Patty made no attempt to cling. She seemed merely to survive, holding herself in a stiff and motionless fetal position. But despite Patty's inability to respond to Lulu's ministrations, Lulu began to care for her child. She examined her as she had when Patty was first born; she turned her over and over, looking at her every part, examining and then licking her clitoris and her anus. She licked the hands and feet. Then she held her daughter close against her soft, warm body. It was a moment of complete calm.

290

Perhaps Lulu felt the muscles begin to relax, for soon she put Patty over her arm. Almost in slow motion, walking on the sides of her feet as she had so many months ago, she began to move about the cage. She glided into Caroline and Joanne's cage and climbed up to the window again. There was not a sound from the baby, who was slung over her mother's arm. Lulu peered out the window, put Patty onto her stomach again, and held her there. Somehow the baby's arms slowly spread out over her mother's hair to grasp her. Patty was clinging.

I began to cry. The tensions that had built up spilled out now in a strange, almost dry weeping. It was as if with that little clinging gesture I knew that everything would be all right, that the weeks of wondering were over, that there was to be no tragedy. They remembered each other. I shook my head impatiently. I could not dwell on my own thoughts. I did not want to miss one gesture, one moment that might escape me.

Carefully and slowly, holding her clinging infant against her with one hand, to be sure of her newfound infant's safety, Lulu lowered herself to the floor. She was handling Patty as if she were a newborn again and as if Lulu were afraid that Patty might fall. There was no way to tell how strong Patty's grasp was, those hands that had not clutched anything but a towel or a blanket for nine weeks. She was not yet clasping with her feet; they still held each other.

Lulu went to the number three cage, paused, then went to the door to look through a peephole. She put the baby down onto the floor next to her; Patty reached out and put her arms around her mother's leg and clung. She no longer crouched terrified or cowered in front of her mother's figure. Perhaps it had been the feel of her mother's body, that touch when Lulu had lifted her, that stirred the memory of comfort in her mother's warmth. Was it, perhaps, the tactile experience that moved something inside her, more powerful than the sight or sound of her mother, more comforting than the kiss, far more primal than any other sense?

The sight of Patty and Lulu together seemed familiar. Things that I had seen so many times in daily life were happening again. Lulu put the baby into her hand as she had so many months before; amazingly, Patty still fitted. Although her arms and legs and parts of her torso overflowed her mother's slender hand, Lulu could carry her like a little football. Lulu ducked under the crossbar and, for the first time, gently swung the baby onto her shoulders. She put Patty down, and the infant clung to her, one small pair of arms encircling Lulu's larger arm trustingly. And when Lulu moved away, Patty followed. Her rear still dragged along the floor in a cowed, fearful position, but she followed. Lulu turned, bent over, and kissed her. Patty stretched out her arms and reached up for her mother. Lulu picked her up. She swung Patty onto her back, and the baby, perched on Lulu's shoulders, grasped the hair around her. Her eyes peeked out at us, just as they had so many times before, wide and shining. Her fear was already dissolving with familiarity. Lulu shifted her onto her stomach, checked her position, and then let go. Patty clung on her own.

There was a whisper of movement behind me and a long, loud intake of breath. Suddenly I was aware of something other than the cages and the gorillas. There was a crowd of people. Their eyes, too, were glued to the sight in front of them. Hardly anyone stirred. Now and again a cameraman changed position, but only to get a better picture. Ronnie was standing on the ledge across the aisle. She was crying. The commissioner, too, had arrived and stood watching the gorilla and her baby. I wondered how long he had been there and if he had seen the reunion.

Everyone seemed stunned by what he or she had witnessed. No one's fantasy could have equaled the tenderness of the reality before us.

Lulu put the baby down, and for the first time Patty lifted her head to look around, already forgetting the fear of only moments ago. She went out on her own to explore the nearly strange cage. She stayed near her mother at first, but soon, as Patty saw Eddie's familiar face beaming at her from the bars, she went to him. For just a moment she looked up into his face and let him stroke her arm. She sat quietly at the bars and peered into his eyes. Then she turned and walked across the cage to Lulu. She put her arms around her mother's leg. At that moment she made her choice.

Suddenly in the silent, nearly breathless atmosphere of that Lion House there was a stir. Dr. Nadler turned to Fitz. "I think it's time to let the male in."

Kongo! I had forgotten Kongo!

292

He was still outside, curious and excited, trying to look in on the cages. Nine weeks earlier he had struggled, agonized in his attempts to see his mate from the same window, from the same cage. He had listened to her screams and was helpless. Today, too, she had screamed and he had been shut away from her. Could he have heard a difference in her screams? Had he been able to get some tiny, instant glance of the baby? Apprehension welled up in me again. In a few seconds we would know if he would accept Patty and if Lulu would have regained the necessary status as a mother, or if Kongo would regard Patty as an intruder and would resent her presence. It was only twenty minutes since Lulu and Patty had met. Now Kongo had to join them.

The door to the home cage was opened, and he strode in, huge and masculine. Lulu moved to the mesh wall and looked through. Then she turned and quickly crossed to Patty, who sat playing calmly with a little chain that swung loose from the bars. Lulu cupped her hand under her daughter's chin and tilted Patty's head up toward her. She looked at her fiercely as if wanting to see what was in Patty's eyes. For a brief moment they gazed intently at each other. Then Eddie pulled the chain between the first two cages. Lulu swept Patty up into her arms and leaped to the crossbar. I could see Kongo's shadow against the screening as he looked in on them.

When Eddie now opened the door between them, Kongo did not enter the cage. He stood in the doorway, looking in on them from a distance. His huge body filled the opening; his eyes were hard and intent. He made no move whatsoever. Lulu put Patty down, picked her up again and, with Patty in her arms, went to him.

At ten fifty-five they came together. They were a family again and complete. Briefly they met and put their arms about each other. Then Lulu simply strolled unconcernedly and casually into the home cage.

We watched with awe as Kongo suddenly and immediately began to perform his duty as a male. He circled his mate and his baby, then strode around the periphery of the cage, stiff-legged and proud. He looked out at the crowd in front of him, his jaw thrust forward, his stance fully that of the responsible, protecting male. He declared his territory and marked off the boundaries. He was guarding his troop. He protected them, and then, only then, did he go to them.

He looked down at Patty, who clung to Lulu's stomach. He stood over them for a moment, then left to circle the cage again and look out at the crowds. His eyes were wide and sharp and had a warning in them. Again he returned to the baby.

Lulu did not allow him to touch Patty. And when he approached them, Lulu picked Patty up and swung up to the crossbar. Finally, Lulu allowed him to stroke his daughter's head. As Patty lay in Lulu's lap, wiggling her arms and legs, Kongo gently pressed two huge fingers into her soft hair. He bent his head down to Patty Cake's with slow and infinite tenderness. Minutes later, after another authoritative stroll around the cage, he put his mouth to her head and kissed her.

We turned to each other in exuberance. The smiles spread over the crowd; the unbelieving whispers grew into rumbles of laughter. How quickly the atmosphere in the Lion House had changed!

Soon the crowds dissolved, and the Lion House was emptied of the equipment. Fitz hunched over the railing before the gorillas' cages, watching them calmly. Dr. Nadler pondered what had just taken place, his hand stroking the bowl of his pipe. Eddie stood in the keeper's room, happily cutting up some fruits, and I stood in my familiar place in front of the cage, taking notes. Kongo strolled over to us, and Fitz began to play with him. Obligingly he pulled the male's fingers, poked his belly, and scratched his back. Fitz understood the importance of Kongo's needs, that they were the natural cravings of the dominant male suddenly thrust into second place. Lulu's involvement was once again divided between his needs and the baby's. And he needed attention. Their play was just as it always had been.

Behind them Lulu and Patty played. Patty was becoming more and more relaxed as each minute passed. She ran to the bars often and then back to her mother. When Lulu tried to carry her along, she sometimes objected and her little sounds of independent annoyance filled the air.

Patty was wonderfully unafraid. She walked around the cage with her little rear high off the ground, not at all the fearful and sub-

294

missive posture of an hour ago. She struggled with her mother and surprised her with sudden scrambles that were stronger and quicker than any Lulu remembered. Patty seemed to be curious, at times eager and even angry, but not afraid.

It was wonderfully familiar. I settled back into the comfortable routine in the Lion House. Had nine weeks actually passed? Had there been a lonely space in these cages, one that had been sad and real? Had Patty and Darwin played together, laughing and giggling as they caught each other in a welcoming, playful hug? Had I ever held Patty Cake in my arms? It seemed as if these things had never been. And where had our apprehensions gone?

From my diary, 4:50 P.M.:

It is late and nearly time for me to go. And now, after all these weeks, the most familiar sight of all, one of the first that I remember seeing all the way back in September: the "family" lying together side by side to rest, their bodies slightly at an angle as they lie head to head.

What a wonderful sight! The sun is pouring in and brushing across Lulu's breast and Patty's head is resting there. Their forms are rich in color. Patty's arms and legs are hanging limply over her mother's body as she slumbers. And Lulu's arm is brushing Kongo's back as he lies next to her, his eyes opened and on the baby. His hand is on Patty's shoulder.

It is so peaceful, and that sad sense of emptiness is gone. The cage seems full and complete again.

EPILOGUE
Everything Changes

Patty's adjustment to life in the cages was astonishingly easy. Were those first six months of her life really so binding, so deeply embedded in her mind? Was the imprinting of her relationship with her parents intense enough and strong enough to have endured her separation from them? Caroline Atkinson deliberately stayed away from the Central Park Zoo for a long time, not wanting Patty to become disturbed at the sight of her. But when she did come for a visit and stood as one among the crowds, Patty barely noticed her. She turned to Lulu instead.

All the questions that we had posed, all the difficulties that we had foreseen and spent so much time pondering, were no problem at all. Lulu readily allowed Patty to come to the bars for her bottles once she and Kongo were introduced to bottles of their own. And once they were accustomed to them, the three of them would sit at the bars, drinking, Patty usually nestled against Lulu's soft stomach, her eyes tilted upward, riveted on Ronnie's face. Lulu's eyes, above, gazed calmly on the same face, and a few feet away someone was playing the bottle game with Kongo, keeping him from finishing his juice too quickly and heading for Lulu and Patty's.

Lulu resumed her role of responsibility toward Patty as if it had never been interrupted. Kongo readily gave up his place of su-

297

preme dominance. Within a day of the reunion Lulu had observed her daughter's new, somewhat spoiled behavior and established a discipline that Patty had not received in human hands. For a while Patty was constantly directed by her mother, pulled, pushed, stopped, held, or merely kept within an area of her mother's choosing. But when Patty finally understood what was expected of her, the disciplines were relaxed, and she was allowed more freedom. She would come when called, go where her mother directed, and she happily scampered around the cages much as she pleased. The gorilla's lives were normal again.

Patty was, in every way, Lulu's girl. Yet as time passed, Kongo watched her with obvious yearning. Always aware of his presence, Patty watched him with open curiosity, until one day he reached out for her. As Patty put her arms around his ankle, Lulu merely glanced their way and casually strolled toward them. And by the time she had scooped Patty up the father and daughter had looked into each other's eyes and touched.

Lulu was far less possessive than she had been before, and eventually there was a casual coming and going among the three gorillas, a weaving and interweaving that threaded through the five-room apartment of their cages. And Patty became Kongo's girl as well.

I loved to watch them. Andrée and I stood in front of the bars as we always had before, and by October what we had hoped for for such a long time had come true: the family. The three gorillas, two mountains and a mouse, hugging and tumbling in play.

And then everything changed.

As Kongo drifted into a more casual attitude toward Lulu and Patty, people who watched became afraid. Kongo was, more than ever, the magnificent monster. He was no longer a mere teenager but was growing, seemingly by the foot, into a blackback male. His massive, powerful body loomed in contrast with Patty's fourteen or fifteen pounds, and as he held her, he seemed to surround her. All his movements seemed to be so much freer and wider than Lulu's, and in his arms the baby seemed helpless and small.

One day in mid-November, as he played with Patty on the platform that Fitz had built in anticipation of Patty's return, he held Patty over the edge, lowered her, then let her drop a foot or so to the floor. He quickly looked at Lulu for her reaction, then reached for Patty again. Patty was furious, clicking and biting, then squirming and wiggling against his hand as he lifted her up to the board. He repeated the performance.

"He's the one who broke her arm,"

"He's going to kill her."

The murmurings in the crowd grew as the keepers ran to report Kongo's behavior to Fitz. Worried and anxious, a small group of us stood in front of the cage and watched closely as Kongo tried to tease Lulu, and Patty scrambled away from him over and over again. Within a day, on November 15, Kongo was living alone.

Two days later Lulu went into estrus. Frustrated, she was rougher with Patty than she had been before, flinging her off when the infant wanted to snuggle, tossing her aside if she got in the way. As soon as the monthly cycle was over, Lulu again became an exemplary mother, but a month later, when she again went into estrus, she became frantic. It was decided that it was too dangerous for Patty to continue to live with Lulu. In mid-December she was tempted away from her mother. The door between them lowered, and she was gone. Within minutes Fitz put Lulu and Kongo together again.

I felt that the separations were unnecessary. I think that there was so much pressure on the administration of the Central Park Zoo from the public, the press, the Bronx Zoo, and the administration of the City Government that the administrators felt that they had no choice. One day in October Patty had twisted her arm in the bars of the cage for an instant. Within minutes someone had called the press, who turned out en masse. It seemed as if they were waiting for something to happen. And now that there was a possibility of an accident, the administration was too tired of the fighting and of the tensions that never seemed to stop. Kongo's behavior with Patty might have been another natural step in the development of a relationship between the male and the infant who was now older, stronger, and better able to withstand greater physical interaction. She might well have begun to assert herself more. Certainly she enjoyed the confusion of their play, and when Lulu and Kongo shuffled and tumbled with each other, Patty had lunged into the melee, eager to join in the fun, giggling in anticipation. Whether the play and Kongo's more casual treatment of her was a preamble to another phase of her development, we would never know.

Lulu's behavior certainly was reminiscent of that she demonstrated during the first month of Patty's life when, without Kongo's presence in the cage, her frustrations mounted and she had left Patty crying in the middle of the cage while she cavorted from place to place. No doubt the fact that she was now in estrus at the same time that she lost her mate increased her frustrations.

There were problems. The problems that existed within those cages were real enough, but under other circumstances they would have been solved without the destructive loneliness that followed for those animals. The solutions, however, called for money. And there was no money for them.

There were problems of space. The five-room apartment allotted to the gorillas had shrunk to three rooms now that the weather was colder and the two large cages outside were closed off. Gorillas, by their size, activity, and nomadic nature, need space—open space that allows the freedom of movement. Without sufficient space, their genetic modes of behavior must become unnaturally modified. There was little space.

There were problems of social interaction. Kongo, the dominant male, was again shoved into the background with no outlet for his energies other than Lulu and Patty Cake. Perhaps another female gorilla could have enlarged this little troop and alleviated the frustration that Kongo felt just as he was growing into his full potential as an adult male. But there was no money and no space for her.

Simple things could have been done to help the problems; things that did not call for a completely new design for a gorilla house, but would have simply entailed shifting the animals around and knocking down walls between existing cages: building more platforms; creating a space where Patty could go and Kongo could not follow. But even these simple things were out of the question; for the city of New York was facing a terrible financial crisis and the money was not forthcoming. Even the manpower was decreasing as keepers were offered jobs in other areas of the Parks Department, where they were needed for other, more immediate work.

The people who cared the most felt trapped in the dilemma, and

300

tired of always fighting, fighting, under the eyes of the public and the city government, they finally succumbed.

The following February the Parks Department administration received a few irate phone calls. There were some visitors who objected to the sight of Kongo and Lulu mating. These few callers changed the animals' lives again. It was the final straw. The word came down: separate the gorillas. The animals would, from now on, live apart.

In the months that followed his separation, as one day led to another, Kongo began to demonstrate various forms of aberrant behavior, until one day I watched, horrified, as he fiddled with his own feces and then put some in his mouth. The weeks went on and on with no interaction, no interest, no spontaneity. He was forced to give up his roles, his genetic behavioral heritage; he had no responsibilities, no controls, nothing but empty space in which to sit. The most that he could do to assert himself was to thrust out at a keeper. There were new keepers who, not knowing Kongo, not being acquainted with his needs, dominated him. Nothing but the separation itself could have been more harmful for him. Kongo was the dominant male dominated, forced into a situation which was contrary to his very nature, forced to be what he was not. Now everything had been taken from him. His frustrations grew, and he became more agressive.

Everything is changed.

Patty Cake has lived with a chimpanzee, Panzee, since her separation from Lulu. She is the image of her father, with the sweet, even temperament of her mother. Her adjustment to life with an aggressive fast-moving chimp was as easy as it could have been. They are pals now, and they both swing toward me, vying for space on the crossbar. They put out their hands for a moment, but I cannot reach them. There is plexiglass between us. Patty's soft eyes gaze calmly into mine. Then, Panzee pulls her leg or pushes her, and they are off, swinging and scuffling in play.

Lulu just sits alone in the small number three cage.

Joanne is dead, and Caroline is alone.

Eddie Rodriguez retired. His arthritis finally became too painful for him to continue such difficult physical labor.

301

Raoul accepted another job somewhere in the park.

Dick Berg is gone, too.

Ronnie Nelson married Nick Arcidiacono, another keeper. They now live on a farm in the far reaches of Wisconsin with their new son.

Only Luis Cerna and Richie Regano are still working in the Lion House. And soon they, too, will retire.

Dr. Garner is no longer associated with the Central Park Zoo or Flower and Fifth Avenue Hospitals, which gave up their interest in zoo animals. He has a private veterinary practice in New York.

Fitz was transferred to another zoo in August, 1974. But now they have brought him back in this January, 1977, when the zoo is in terrible decline. If anyone can help, he can.

It breaks my heart to go there now. For the cages finally seem to be really as empty as they look. Without the life that those animals gave to one another the cages can only be cement and stone, tile and steel. The animals have reverted to being museum pieces that just happen to breathe for all those people who pass by and do not know them or what occurred in the same cages once, not too long ago.

Kongo has become aggressive, they tell me. But when I go to see him, he comes to the bars and puts his two fingers out for his lollipop. His eyes are soft. He wants me to stay. And when I leave, he watches after me and sometimes pounds the thick tile wall just as he used to do. It trembles. The sounds reverberate through the Lion House as if a tremendous burst of thunder were crashing and echoing through mountains. And I hear an anguished, forsaken loneliness. I know that this magnificent dominant silverback male gorilla is helpless, caged, with no choice but to exist in a vacuum that will have no end until he dies.

I remember a day when I gave Kongo his bottle and Luis warned me, "Don't let him bite off the nipple!" But Kongo made no attempt to bite off the nipple or snatch the bottle away from me. He stood at the bars, his hands folded under his wrists, his eyes softly gazing at me. I held the bottle for him and squirted the milk into his open mouth. I could see his gums and his huge tongue sprinkled white with the milk. Then, as he watched me, I put it into his mouth and he sucked on the nipple. His hold was so powerful that even through his peaceful, docile feeding I could feel his immense power. When I took the bottle away, he did not try to keep it. At my slightest indication he let the bottle go.

Suddenly Lulu came up behind him and slapped him on the back. He turned, put his hand on her head, and bent down to the baby. Patty Cake peeked up at him with round, shining eyes. Then the three of them walked away from me to the back of the cage. Together.